TRAINING DESIGN IN AVIATION

Training Design in Aviation

NORMAN MACLEOD

Routledge
Taylor & Francis Group

LONDON AND NEW YORK

First published 2001 by Ashgate Publishing

2 Park Square, Milton Park, Abingdon, Oxon OX14 4RN
711 Third Avenue, New York, NY 10017, USA

Routledge is an imprint of the Taylor & Francis Group, an informa business

First issued in paperback 2016

British Library Cataloguing in Publication Data
MacLeod, Norman
   Training design in aviation
   1.Aeronautics, Commercial - Employees - Training of
   I. Title
   387.7'07155

Library of Congress Control Number: 2001086759

ISBN 978-0-291-39844-4 (hbk)
ISBN 978-1-138-27527-0 (pbk)

Transferred to Digital Printing 2014

Contents

List of Figures and Tables

Preface

This book has been written after a career spent floundering around in an attempt to design and deliver effective training. It should be read by anyone about to embark upon either the design of a new course or the updating of an existing course. Those managers responsible for the training process should also read it. We have tried to encapsulate the key issues around the process of training design which need to be controlled if our workforce is to perform at anywhere near their best once they have been through our course. We have often felt that many graduates achieve outstanding results despite of rather than because of the course they have been forced to endure. From this comment the reader may assume that the approach adopted in the book is idiosyncratic, even prejudiced. If so it is only in an attempt to throw existing training practices into some sort of relief.

We have tried to illustrate the issues in training design through examples from a range of projects we have been involved in over the years. The approach is heavily pilot-centred and also based around the training domain of Crew Resource Management. This is a simple reflection of the author's focus for the past few years. Readers should look past the surface features of the situations and concentrate on the underlying performance problems. We could also be seen to be unfairly critical of the efforts of others. This has never been our intention. Much good work is being done through the sheer initiative and creativity of those involved in providing training in airlines. How much more could be achieved with a little structured thinking?

1 Introduction

Every day, somewhere in the world, events occur which could have been prevented if the participants had been better trained. Take these examples, for instance. In October 1994, a Boeing 747 took off from Sydney Airport bound for Japan.[1] An hour after take-off an engine had to be shut down and the crew decided to return to Sydney. On final approach, the aircraft's nose-wheel failed to lower and the aircraft touched down with no nose gear. The investigation revealed deficiencies in the competence of the Flight Engineer as well as serious mismanagement of the crew training associated with the introduction the aircraft into service. The flight was the inaugural long-haul route for the company concerned, a factor which contributed to the situation.

In September 1997 a member of the cabin crew aboard an MD-80 aircraft was approached by a ground handling agent who asked about a toilet door that needed fixing.[2] The in-bound crew had reported the door faulty but, having just boarded the aircraft herself, she was unaware of the problem. The agent talked about the toilet in the First Class section needing some WD-40 lubricant. 'They didn't have WD-40', he said 'but I found something just as good'. The job done, the cabin crew member commented on the strong smell and the agent replied that he 'had only used a little'. After take-off, 2 of the cabin crew and one passenger fell ill and needed hospitalisation, which resulted in the aircraft being diverted. The passenger handling agent was an office manager, not a technician, and his substitute for WD-40 was a highly-toxic lubricant which was not to be used in confined spaces, a fact clearly marked on the container.

In June 1998 a 747 taking off from San Francisco, again bound for Japan, experienced problems with an engine soon after rotation.[3] The First

Officer (FO), who was flying the aircraft, applied the wrong technique to keep the aircraft under control. The aircraft lost speed sufficient for the stall warning stick shaker to be activated. The aircraft cleared San Bruno Mountain, 5 miles northeast of the airfield, by 100 feet, missing the radio masts rising a further 600 feet above the mountain top. The subsequent investigation found that most FOs rarely handled the controls on take-off and landing. In fact, some were flying for up to 2 years without touching the controls during these critical stages of flight, other than in the simulator. Furthermore, the airline discovered that FOs were regularly letting their currency lapse in order to gain extra holiday, given that they could not fly between losing currency and completing a simulator re-qualification.

These 3 case studies may raise a few eyebrows but each highlights a problem which most aviation professionals encounter everyday. In order to keep the schedule operating we have to be flexible. So, the office manager who picked up a can of lubricant was only trying to help. That said, we need to consider the way in which the company identified the skills needed of its workers. Perhaps a qualified engineer was not required but we can examine the approach to health and safety training, product knowledge and the identification of hazardous materials. The Flight Engineer on the 747 had a history of training problems but he was not helped by the situation he found himself in. Inter-airline rivalry and over-stretched management degraded the training intended prior to the introduction of the 747. Now commercial pressures have intervened although selection methods, course graduation standards and overall project management are all implicated.

The final example, again involving a long-haul aircraft, reveals some of the problems associated with maintaining skills in an environment that offers little opportunity to put training into practice. The last example also shows up another aspect of aviation that will be addressed in Chapter 1. Aviation, like all economic activity, involves individuals, all of whom have their own agenda. The FOs in this airline had developed a scheme to augment their days off, at the company's expense. Work can be seen as a constant struggle to balance a set of diverse needs, including the motivation of the workforce. This struggle can have an impact on training.

Ironically, training has a long tradition in aviation. Such was the difficulty of mastering early flying machines that the need to train was well-established from the beginning. Aviation training devices have been at the forefront of technology; solid-state electronics were widely used in the design of flight simulators long before they were commonly found in aircraft. Add to this the regulatory requirements which mandate a need for regular refresher training and we have a climate in which the need to train is unquestioned. However, that is not to say that all training in the aviation community is high quality, or that all those in need of training get what they require. Rising accident rates, increasing pressure on budgets, deregulation and privatisation are all forcing a re-examination of the time and resources devoted to training in the industry. And despite the benevolent attitude alluded to earlier, it is probably fair to say that many aircraft operators only do the absolute minimum training necessary in order to remain legal, and they probably do not even do that very well.

With the spread of new technology on the flight-deck, the increasing emphasis on revenue management, the inexorable drive to extract even more work out of reduced staff numbers, it is timely that we re-examine the role of training design in meeting some of these operational goals.

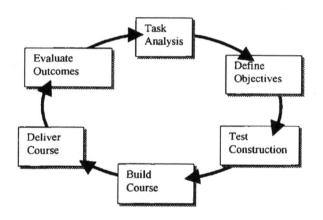

Figure 1.1
The Training Design Cycle

This book is aimed at airline personnel who have a responsibility for the design and management of training. It assumes that the reader will have some experience of delivering training. It is not intended as a guide to the production of classroom materials, there are better books available on developing the various types of training material. Equally, this book is not about how to deliver training in a classroom. The main aim is to broaden the reader's understanding of training as an activity and some of the problems associated with training design generally. We will discuss methods and techniques and offer suggestions as to how to solve training design problems. However, our main objective is to reflect an idea of what constitutes best practice in terms of training design.

The book is arranged along the lines of a classical model of training design with a chapter devoted to each stage of the model. Our intention is to broaden managers' and training designers' understanding of the process. We will discuss the key issues surrounding each topic and suggest methods to be used at each stage. We will highlight weaknesses in current approaches to training and, with one eye on the future, outline strategies that will prevent the reoccurrence of the incidents with which we opened this Introduction. We end each chapter with a short exercise which will help you to consolidate what you have learnt.

2 Organisational Context

Introduction

Training design is, quite simply, the identification of a desired workplace performance which can be met by some form of training intervention and, subsequently, the design of a solution which can efficiently and effectively meet that need. Our aim must be to provide some form of guarantee of success but at an affordable cost; conversely, we cannot afford to waste money or run the risk of producing incompetent workers. Implicit in this goal is the assumption that training is the answer. This may not always be the case. Training is only one of many organisational processes; as we will discuss later, poor performance could reflect management practices, resource allocation or poor selection of personnel in the first place. Therefore, although our focus is clearly on performance in the workplace, we want to start this examination of training design practices in aviation by looking at the organisational context within which the activity takes place. At the end of this chapter you will:

- Have considered the economic, political, social and psychological factors that can influence training decisions.
- Have considered strategic reasons for poor workplace performance.
- Have considered a stakeholder analysis of training decision-making.

How to Guarantee Failure

As an introduction to this discussion of the organisational context, let us consider an example of how easily things can go wrong. A British airline invested in a ground breaking interactive video-based course designed to tackle management problems on the ramp. The hardware chosen as the platform for delivery was incompatible with the company's corporate Information Technology (IT) policy. In fact, when the project started, the company did not have a policy. By the time the interactive video course was complete, a policy had been agreed which had settled upon PC-based hardware as opposed to the AppleMac used in the project. Thus, from the outset there was no chance of maintenance or spares. Change processes in other parts of the company had an indirect impact on training strategy. So why didn't they redesign the package to run on acceptable hardware? In the early days of technology-based training, authoring packages were simply not exportable across different computer platforms. However, at much the same time the training department went through a reorganisation and the manager responsible for the project found himself out of favour with the new department head. The eventual outcome was that the training package was consigned to a cupboard. So, although an excellent solution to a very real and serious problem got several airings at conferences and exhibitions, its use in-company was short-lived.

This simple tale illustrates two of the dimensions by which we can define the organisational context; economics, in the sense of investment and pay back, and politics because training interventions can sometimes challenge perceptions of who is in control of decision-making. In addition to these, in this opening Chapter we will also need to consider the sociology and psychology of organisations as these will also influence any training solution we devise. Of course, none of these factors work in isolation within the workplace; in the real world, these four dimensions will be inter-linked. Nor are they present in every situation, after all, some training decisions are simple and straightforward; we should not overplay the hand at this stage! However, we have used them as a starting point because they are neatly summarise some of the key problems which have to be resolved before you even get down to the task in hand, which is to design training. Let us look at them in more detail, starting with economics.

The Economic Dimension

In the example we have just seen, the need of a large organisation to look for economies of scale by standardising on the computer technology used within the business outweighed the benefits of supporting a non-standard system just for one training application. The point here is that training has direct economic effects; it involves expenditure. More important, expenditure on training is usually seen in a negative context. It is difficult to show any direct return on the investment. Although the need for basic training to make staff productive is readily accepted, the scope of, say, induction training for improving staff retention and morale is rarely explored. To reinforce this failure to link training with a company's bottom line performance we need only cite the example of an airline which, once the monthly figures got too bad, shut down its training centre in an attempt to save money.

We can identify some of the direct economic inputs, such as: training staff costs, materials purchased, trainee costs, replacement costs in the workplace to make up for staff in training (or productivity lost as a result). There are also indirect effects. For example, raised skill levels may result in different expectations within the work force which, in turn, may require different reward schemes. We can, of course, identify some positive economic benefits of training, such as reduced error, faster turn-around, employment of new technology or working practices. It may be that more part-time workers can be employed or greater use made of outsourcing.

It is impossible, then, to think of training without also thinking of the costs involved. That said, surprisingly few airlines have budgets for training and fewer still actually know how much the activity costs them. One of the problems is that training often gets hidden. For example, some cargo or charter operators use empty positioning sectors to do pilot recurrent training. Given that the crew were required to fly the aircraft into position in the first place, the training is deemed to be free. Quite often, the scarcity of approved simulators for a specific aircraft type reinforces the view that it is cheaper to train on the aircraft. One UK operator of regional/commuter-sized aircraft ships its pilots to the US for refresher training in a simulator. The bill in 1998, given that an instructor from the airline was almost permanently based abroad, came to about half a million pounds. Again, some charter companies use line pilots to conduct Crew Resource Management (CRM) training during the off-peak season. Because there is no alternative use for these staff, the training is once more considered to be at no cost. In some cases, instructors prevail upon colleagues to give lectures as a favour. Quite often this is rationalised in terms of some perceived added benefit. For example, we know of some cabin safety trainers who had a pilot colleague deliver a short presentation on the theory of flight to new-hire cabin crew. Whilst laudable and, no doubt, of real benefit, it still represents unmanaged and uncosted training.

The real irony in all this is that, despite the fact that, in full-flight simulators, airlines employ some of the most expensive training devices in the world, training for staff other than flight-deck personnel is considerably under-resourced. Andersen Consulting,[4] in a 1995 study reviewing airline training in the United States, found that carriers spent, on average, US$ 281 per year on cabin crew and US$ 3,565 on flight-deck training. Figures for a wide range of other US industries showed an average annual expenditure of US$ 800 - 1000 per employee. Using figures from the UK Civil Aviation Authority, again for 1995, we see that UK carriers fare little better. Whereas pilots attracted an expenditure of £3108 ($4972) per year, cabin crew training cost airlines from as little as £198 ($316) per head to as much as £1616 ($2585) per head. For comparison, UK industry as a whole spent an average of £384 ($614) per employee on training. An interesting point is the fact that scheduled airlines spent nearly three times as much, on average, on cabin crew training as the charter operators did. Cabin crew training costs for a range of UK airlines are given in the Table 2.1 below.

Table 2.1
Representative Costs of Cabin Crew Training for a range of UK Airlines (FY95/96)

Scheduled Operator		Charter Operator	
Number of Cabin Crew	Training cost per person	Number of Cabin Crew	Training cost per person
575	625	911	198
853	1489	935	1310
1399	1616	1240	222
10454	1106	1460	191

Just to reinforce this attitude to training costs, we were once asked to provide instructor training for a small airline to allow staff to run an internal 2-day course. Management's view was that if the course was 2 days long then you only needed 2 days in order to learn how to teach the course. Readers will see the flaw in this logic but will not be surprised to learn that we did not get the contract. On a larger scale, when the question of training maintenance personnel in human factors was raised in one country, it was decided that the industry was unlikely to accept the need to take engineers off the line to attend courses. As a result, the preferred solution was to focus training on the initial licensing stage.

Having looked at some aspects of the economics of training, we will now move on to consider politics, or the exercise of influence.

The Political Dimension

Economic factors are readily identifiable and, as we said earlier, are often the first to be considered when the question of training is raised. But we also saw that political influence, albeit at the level of office politics in the case of the ramp training interactive video, cannot be ignored. At the other extreme, the initial impasse, which was a catalyst in the B-747 mishap at Sydney, was only resolved when the Chief Executives of the 2 airlines involved sat together around a table.

The reasons for a training requirement may not be rooted in some clearly identified need. For example, the introduction of Crew Resource Management (CRM) as an item of recurrent training within the United Kingdom caused considerable confusion in a large section of the industry. Air Taxi companies could see little application to their own particular single-pilot mode of operation and yet were required to study the same topics as their multi-crew colleagues. The Regulator insisted and companies had to comply. But what was the motivation for introducing the training in the first place? The fact that more aircraft accidents are the result of human rather than technical failings is now received wisdom and yet our understanding of the exact cause of an accident is often incomplete. The Regulator is charged with the safe conduct of aviation, itself a political act, and so had to do something. The result was a vague solution to a vague problem.

Of course, it is not unknown for higher management to act on a whim but we can see the difficulties caused by ill-defined requirements if we look at the way CRM training was applied across the industry. The CAA's requirements were framed around the major air carriers who could afford to release crews for training. However, air taxi operators argued that time on a course represented, for them, a loss of earnings, given that many of these companies were run with a single pilot. Thus, CRM training requirements were set at 3 days for air carriers and 1 day for air taxi operators despite the fact that both groups were meeting the same regulation.

We underestimate the politics of training at our peril. Consider this; pilot unions in many countries have insisted that CRM training is de-identified. Thus, if an individual performs to an unacceptable standard, the employer can take no overt action for fear of union reaction; the irony being that all pilots accept that those who need the training most are least likely to accept the message. At best, airlines are presented with averaged data of the performance of the work force as a whole. In a similar way, initiatives such as the Advanced Qualification Program (AQP), in America, and various company confidential incident reporting systems are hedged about by immunity and anonymity. Of course, the reality is that poor performers are scheduled for 'additional training'. However, not every country has a legal system which would allow the blame-free environment needed for full and frank disclosure of errors in the workplace. To make matters worse, just as the industry is starting to appreciate that this protectionist attitude is becoming outdated, the increased risk of protracted law suits over liability claims after accidents is causing some to question the wisdom of too much recording of performance. In Australasia, threats by police to subpoena flight and cockpit voice recorder data after a crash has led to fears of pilots being prosecuted for transgressions and the subsequent suggestion that pilots will unilaterally turn off recording equipment in order to protect themselves. Our concern here is not to argue the rights and wrongs of this situation but simply to recognise that training must be seen in a broader, and changing, political context if it is to be effective.

Even though there is a clearly identified training need, it may not be formally recognised because the organisation has not got the resources to act and so prefers to ignore it. Sometimes, recognition of need implies a shortfall in performance which may be politically unacceptable to admit. Recognition of a training need may require a restructuring of the organisation, with the possible loss of senior management jobs. In which case we find that the desire to maintain existing organisational structures outweighs the requirement to train. For example, as the 20th Century drew to a close, it became increasingly clear that the organisation within Korean Air was flawed. Its accident rate was unacceptable and its management culture suspect. An attempt to form an alliance with the American carrier Delta had stalled. Part of the problem seemed to be the way the airline was used as an outlet for ex-military pilots who lacked the skills required for multi-crew commercial aviation. Although the airline is under scrutiny as this is being written, and explanations for Korean's poor performance can only be conjecture at this stage, the fact remains that political pressures within the company have maintained a situation which has become increasingly untenable. At the same time, those pressures have undoubtedly prevented training measures being introduced which may have gone some way to moderating the situation the company finds itself in.

Politics and can overlap. In the United States, applicants for some pilot jobs will only be interviewed if they already hold a type rating on the aircraft used by their prospective employer. In effect, the burden of initial training is put on the individual. Is this the action of a responsible employer that it relinquishes control over the entry standards of its pilots? Applicants, presumably, are free to find training wherever they can. Because it is coming out of their own pocket, and without the guarantee of a job at the end, it would not be surprising if individuals went to the cheapest source of training. What control is there over the standard of provision? If we return to the development of CRM in the UK, within 2 years of the regulations coming into force, over 20 individual CRM courses were commercially available within the UK, some of which the Civil Aviation Authority publicly stated as being substandard in either content or method of delivery. In both cases, the product offered for sale, be it a type rating or a CRM course, is mandated as acceptable by the regulatory authorities and yet the value of the product to the industry seems dubious. However, probably because decisions by purchasers have been based on cost alone, the level of proficiency achieved by users of the courses is possibly less than we would like; but it is legal.

The Sociological Dimension

Having looked at the economic and political dimensions, we now need to consider the sociological framework within which work occurs. Aviation is a system of production, like any other industry, in that aircraft, people and a complex infrastructure are used to create wealth. The aircraft is a tool within this system. In the same way, the aviation industry responds to global economic cycles, like any other industry, which results in changing working conditions that can influence peoples' performance. One of the business models gaining dominance at the turn of the Millenium was that of outsourcing. The poor administrator implicated in the toilet door story in the Introduction worked not just for a sub-contractor (all this airlines maintenance was sub-contracted) but for a sub-sub contractor. In effect, a business strips out activities that are not core to its success and finds an alternative supplier. By then forcing others to compete for this ancillary work, costs are driven down. Airlines become increasingly smaller entities embedded in a system of suppliers until we get to the virtual airline which operates leased aircraft, flown by contract crews, supported by service suppliers; only the Chief Executive is on the company payroll. Fanciful, perhaps, but the model of Southwest Airlines is being adapted in many countries. However, what works well in one set of circumstances may not always transplant to another environment. What we are interested in, at this stage, is the impact of this change in the process of work on training.

Let us look at a more specific example to illustrate the point. Another buzz-word of the mid-1990s was 'empowerment'. As a concept, it can have a very real effect on work-place performance. In a study of operations at a European airport, a common theme which emerged from interviews with pilots, cabin crew, ramp and gate agents was that, all too often, a problem could be identified which needed solving if the aircraft was to get away on time. However, in many cases, it was not clear whose responsibility it was to take action and where boundaries of responsibility lay. Advocates of empowerment would argue that everyone in the system has the responsibility and authority to take whatever action was appropriate. Empowerment can be seen as an attempt to get staff to walk towards problems, not away from them. In less prosaic terms, responsibility is delegated to those responsible for carrying out the work. By so doing, management gurus would have us believe that workers will show greater commitment to the task, flexibility, imagination and so on. What is clear, though, is that empowerment can cause a change in relationships. This is particularly true when we consider the organisational changes discussed earlier. Whereas, once, everyone wore the same uniform - at least at the main base - and, therefore, it was possible to tell who should have been responsible for taking action, under the regime of sub-contracting, it is almost impossible to tell whose job it is to fix a problem. Therefore, successful 'empowerment' training will need to address these social changes as well as any specific messages more narrowly associated with making people 'empowered'.

Changing skill levels within traditional workplace groups can give rise to new sources of conflict between line and management. Developing new skills can cause changes in the ability to exert influence. For example, deregulation and increased competition have increased the power of non-flying staff in airlines as the economics of the real world reinforce the need for airlines to make money. The old truism that no profit-making airlines are run by pilots is increasingly becoming a reality as aviation moves from a predominantly public-sector mode of operation to business models based on private sector operations.

Again, we can see interrelationships between dimensions. For example, the interplay between the sociological and the political dimensions is neatly demonstrated by pilots' perceptions of CRM training. Although they are clearly the object of study, in that CRM looks at why they do the things they do, they are also the subject of planned change because CRM is intended to make them less error-prone (amongst other things). However, pilots also feel that they have little influence over the shape and conduct of training and that they are not fully represented in decision-making about CRM. As a result, its is true to say that a significant number of pilots feel resentment towards a well-intentioned training initiative.

The Psychological Dimension

Finally, we come to the psychological dimension which, perhaps, is the primary interpretation of training as an activity. However, we are also concerned, at this stage, with the social psychology of organisations. For example, the airline involved in the Sydney B-747 incident had experience of the B-767 aircraft. On this type the role of the Flight Engineer (FE) was one of monitoring systems. With the introduction of the B-747, a new way of thinking about the FE was required. However, conditioned by previous thinking, the airline failed to provide adequate training for the FEs on the new aircraft.

Training should result in a change in ways of thinking or in ways of doing, and the rest of this book will look at that aspect in more detail. However, in much the same way as training can change the social dynamic of the workplace, so too can training change the way the individual views work. Improved performance in one aspect may generate new ways of looking at other aspects of the job. In some cases, this may be considered a bonus. However, there could be a downside. There is increasing interest in Europe in degree-standard pilot training, something that has a long history in the States, Australia and New Zealand. However, some charter operators who are customers of a UK scheme which places low-hours licence holders, who are also university graduates, as First Officers have not been convinced.   They have found that flying night runs to Mediterranean resorts, with full loads of holiday makers, is enough to convince some prospective pilots that they, perhaps, have made the wrong career choice. That is, perhaps, not a bad thing. However, the point is more one of the impact of high levels of training on job satisfaction and the need to recognise that highly-trained people tend to have different motivations and, possibly, higher aspirations. At this point we intersect all 4 of the organisational dimensions. Increased training leads to changes in people that lead to changes in the workplace, which upsets the political balance and generates a cost.

Is Training the Answer?

So far, we have tried to sketch out the main strategic issues that you need to have in mind as you set about assessing training requirements in your organisations. First, though, there is one more question you should ask before you get too far into the mire; is training the answer? Training is all about optimising performance. We want people to be able to perform efficiently, either through their own efforts or in terms of their ability to exploit the tools of the trade, so to speak. Quite often, poor performance is the consequence of inadequate investment. The people can do the job but have the wrong tools. For example, inspecting aircraft wings for the presence of ice is a difficult and time-consuming job to do properly, and yet, if not done properly can lead to the aircraft failing to get airborne. An examination of de-icing procedures found several reasons for poor performance. The lack of adequate step-ladders to enable technicians to easily and safely get to areas likely to be affected; suitable probes that would allow technicians to reach iced-up wing areas did not exist; the sticky de-icing fluid sprayed on the fuselage obscured the view from windows which further prevented observation of affected areas. Although training was seen as a way to improve the performance of de-icer crews, inadequate investment in tools was probably a more significant problem.

It may be that work is inefficient because it is poorly organised. There have been many reports of poor communication between flight deck and cabin crew resulting from the 2 groups working for different parts of the organisation. Although the aircraft captain is vested with the authority to operate the flight on behalf of the company, the commercial needs of keeping the customer down the back of the aircraft satisfied can lead to disagreements about how best to solve a problem. For example, pilots will often try to make up time on a flight, thinking that passengers are always pleased to arrive early at the destination. Unfortunately, cabin crews need a finite amount of time to deliver the service. Given that they are often also under pressure to meet targets for sales of duty-free goods, any shortening of the flight time can have severe repercussions. Failure to make the duty-free round can hit cabin crew in the pocket as a result of lost commission whereas the pilots' salaries are safe, no matter how quickly they get to the destination. Some airlines are bringing flight deck and cabin crew together under a single head and, hopefully, business goals can be better co-ordinated, leading to more efficient working.

Finally, it may be that people simply have not been prepared for work and so training is the answer. Unfortunately, training can be seen as a cure for many ills, some of which are not amenable to a training treatment.

Levels of Analysis - Stakeholders in Training

The discussion, so far, has reflected a view of work in which many different parties, or stakeholders, all have an interest. At times, interests may conflict. If we consider the following list as possible stakeholders in some of the problems we will be dealing with, we can see that a single solution that will satisfy all concerned will be difficult to achieve:

- Regulator or Industry-level bodies
- Company
- Section/Fleet
- Work group/Crew
- Individual
- Customer/Passenger

This list is also an indication of the levels in an organisation at which training needs occur in one form or another. It is probably fair to say that most emphasis is placed on training the individual level, with some attention directed towards team training. Rarely does the spotlight move any higher up the list. However, consider this problem; an airline has experienced a number of incidents in which pilots failed to follow Standard Operating Procedures (SOPs). On one occasion, an aircraft came close to being involved in a serious accident. Management is convinced that, although the pilots conform to the rules during periodic training in the simulator, there is widespread non-compliance with SOPs out in the real world. The solution is to run a short course reminding the pilots of the importance of adhering to SOPs - or is it? It seems that, on investigation, the attitude and management style of senior fleet managers established a climate in which a cavalier, nothing-is-a-problem attitude was considered to be the requirement amongst line pilots. The SOPs were fairly well defined but they were not all-encompassing and some of them were not particularly well-explained. In some cases, there were many different ways of operating the aircraft and the SOP seemed to have no advantage over any one of a number of different solutions.

At the level of the individual, there did seem to be a need to better explain the rationale of some of the procedures. There also seemed to be a need to work on attitudes towards adherence to SOPs. This last point could be carried through to team training. There did not seem to be a culture of questioning non-adherence - or, at least, accepting those occasions when, usually, the captain chose to make things up as he went along. At a corporate level, there seems to be a need to work on establishing the right culture and confirming the role of SOPs in the company operation.

There is another, more compelling, reason to look more broadly at the organisation when trying to identify training solutions. We have already, possibly flippantly, referred to 'empowerment' and saw how one airport organisation seemed to be suffering from a lack of it. Most of training is involved with getting people to either do things well or do things better. Future training will be about getting people to do things a different way. Thus, if I am sufficiently empowered, whenever I encounter a problem at work, I will look for alternative solutions, even creating new ways of getting the job done. In so doing, I will be looking to gain competitive advantage over other airlines or providers of support services.

In order to achieve this goal, the whole organisation will need to learn. At the airport we mentioned earlier, management decided that on-time departures from the gates were a priority task. A team of 'chasers' was established who wandered the airport with a mobile phone descending on any gate where a delay looked as though it was developing. All other airport personnel wore uniforms whereas these people were in suits. The chasers seemed to have no function other than to report back to senior management on progress, moment by moment. They contributed little or nothing to solving problems and only antagonised hard-pressed staff, most of who were well-able to deal with matters as they arose. However, the sight of one of the chasers sweeping down the jet-way to the aircraft door was enough to take the edge off of the best of days. In terms of dealing with a very real problem, the higher-management solution was not very effective. In fact, what it did reveal was that management needed to learn more about what happened on the ground. The initial intention of improving efficiency and, hence, profitability at the airport was sound. The point of entry into the problem was flawed.

It may help, at this stage, to draw up a list of stakeholders who will have an interest in your training design project. You may want to involve others in a little brainstorming in order to make sure that you have not missed anyone. For example, we were once involved in the complete redesign of a leadership development course which made extensive use of outdoor activities as part of the programme. Having sought the views of everyone concerned we set to work on putting the course together. When we came to plan our first demonstration of the new product we discovered that the one stakeholder we failed to consult was the transport office that would be providing vehicles for our use. The sequence of our activities was incompatible with their other tasking and they certainly were not going to buy an additional vehicle just for our occasional use. So, back to the drawing board!

Having identified the stakeholders, try to answer each of the follow questions. You need answers for each stakeholder.

- What do they want out of the project?
- What can they give to the project?
- What does the design team need from them?
- How can we help them?
- What contingency plan do I need should they prove difficult?

From the answers to these questions you can then draw up a list of organisational factors which are relevant to the project. Each of these factors can be given a weighting, say, between 1 and 10. They can also be identified as positive or negative factors: pushers or pullers. The factors can be portrayed graphically as in Figure 2.1. In the diagram, the length of the line represents factor weightings. Once constructed, the chart can be used to identify project management activities that you may need to incorporate into your overall design. For example, you may need to make an extra effort to keep certain individuals fully informed about progress or you may need to delegate certain parts of the project to a particular person. The object of this exercise is to identify those factors that will have an effect on the design of your proposed training solution.

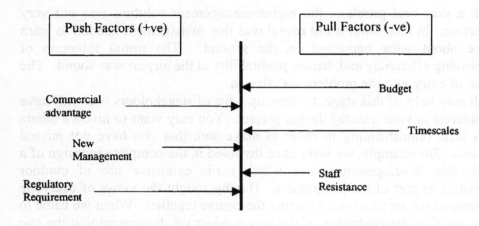

Figure 2.1
Force Fields

Some Questions of Training Strategy

Having mapped out some of the strategic aspects of training design, including the role of stakeholders, we need to look at some of the issues that will be affected by the way your company works. We have not yet started to identify problems that can be solved through training and we must be on our guard against identifying solutions first and then look for the problem afterwards. As an example of what we mean, we were involved in what was probably the second attempt to develop a computer-based training solution for CRM. When initially presented with what we considered to be a bizarre proposal, especially at that stage in the development of CRM as a topic, we found that this was to be a part of a much bigger project. The main project was now coming to an end and the equivalent of 20 hours of courseware remained to be filled. Someone suggested CRM and the rest, as they say, is history.

It does not hurt, then, to map out the boundaries which will constrain our freedom of action. The areas likely to be affected by strategic decisions are:

- Training methodology
- Training delivery
- Design of materials
- Resources available
- Culture

The training methodology deals with the broad framework within which training will be delivered. For example, we may design a full-time residential course or a series of modules linked by periods of work experience. Training delivery covers the actual method of imparting instruction; instructor-led events, video, computer-based training. The materials we use could be specifically designed for the purpose or may be bought in. So, we could use a standard reference to support our aerodynamics lessons or we could produce a handout. The book may not exactly match what we teach and would probably include superfluous material. The handout may be better matched to our teaching but would require effort to produce and to maintain. We may not have access to resources that could contribute to more effective instruction. For example, when we discuss visual illusions in the aviation medicine course, a demonstration of those illusions would be more compelling than an instructor's description. However, not many schools have the resources to lay on such a demonstration. Finally, the culture of the training organisation will have an impact on the students. Sponsored students may be required to show maximum progress within a minimum period of time. Sponsors, after all, do not want to waste money and nor do they want to support students who have a high risk of failure. Under these circumstances, the need for check-points, periodic reviews, examinations etc. will influence the culture within the training organisation and will, in turn, generate a high degree of stress within the system.

Conclusion

In this chapter we have tried to sketch out a background to the training design process. Like most things in life, training is not always as simple as it perhaps looks. To ignore the indicators that training is needed and to be blind to the effects training will have will guarantee failure. Similarly, to fail to recognise that others may not see things the same way as you will result in disappointment all round. The organisational dimension needs to be explored before embarking upon a training design venture.

Case Study

In the next chapter we will consider what we mean by training in more detail. First, though, we want you to consider this problem. You have been asked to advise the Chief Pilot of a small regional airline. He tells you that he is having a problem with the customer services staff who work at the check-in desks. Because they are not doing their job properly, aircraft are late off or are improperly loaded. Baggage is being left behind or put on the wrong aircraft. He wants you to design a course to improve efficiency. You start by wandering around the terminal building to see how things work. Draw up a framework of the organisational aspects you would want to investigate before you tackle the training requirements of the check-in staff.

3 What is Learning?

Introduction

In the last chapter we looked at the organisational context in which training takes place. We tried to identify some of the strategic issues that will need to be addressed if our attempts to shape training are to succeed. Training is all about bringing about change. People who previously could not do something now can perform to the required standard as a result of some planned intervention and, therefore, it follows that learning has occurred. Learning is one of those words that we take for granted and yet the process of learning is still poorly understood.

The advent of brain scanning techniques is only now allowing us to monitor brain activity whilst people do a job of work. The technique is throwing new light on the process of learning at the level of neurones and synapses but we are still some way off a complete theory of learning at the molecular level.

In this chapter we want to review some current thinking about what we mean by learning. As training designers working in an economic system, we need to ensure that the money spent on training is not wasted. Therefore, it makes sense to see what guidance we can get from the way students learn in order to make our training interventions more effective. At the end of this chapter you will have:

- Considered the differences between education and training.
- Considered how differences between individuals affect learning.
- Considered how learning occurs.

• Considered how we develop expertise.

Training and Education

Traditionally, handbooks like this one have tried to differentiate between training and education. Training tends to be tightly focused, it deals with the specific needs of the job, whereas education is more broad-based. One concentrates on work skills while the other is more generalist. One deals with short-term results while the other is part of a long-term developmental process. The different concepts are well-illustrated by a story about parents who, although concerned when their child came home and said that sex education was being introduced to the curriculum, became positively irate when the child went on to say that the next stage would involve sex training. You get the idea.

We can look at a more relevant example if we return to degree-based pilot training courses mentioned in the previous chapter. The former Soviet Union had a university-based pilot training scheme which took 4-5 years to produce a pilot qualified to fly as a First Officer on a single piston-engine aircraft in daylight and clear of cloud. But it also produced a highly qualified engineering graduate who was guaranteed a minimum wage within the social structure that prevailed at the time. University-based aviation courses in the USA have a long history but the rapid rise in their popularity, mirrored in Canada and Australasia, is a reflection of the view that a university degree is increasingly becoming a pre-requisite for entry into the job market. In the UK, the opposite view is more commonly held. Pilot training takes the shortest time deemed acceptable and pilots emerge with just a licence and no other qualification. The syllabus reflects what a pilot needs to know and to do to be safe in the job.

Here, two systems have the same end result but one takes an educational route while the other is an example of a system largely based on a training model. Both approaches, hopefully, result in a change in state. The person subjected to either process is changed in some way and, as we have already said, if the change is in the right direction then our job as training designers has been done.

Do we need to reconcile these approaches? As the concept of change in the work place becomes prevalent there is recognition that life-long learning will be needed in order for people to remain competitive and employable. Thus, training programmes, delivered at the point of need, which broaden and provide transferable skills will prevail. The implication being that minimalist, one-shot training delivered on entering employment will have a limited future.

So what can we learn about learning from this examination of the training/education dichotomy? Probably the main observation is that the two approaches seem to differ in the depth of instruction given. After all, if you are going to spend years studying the same course that others cover in months then there has got to be something different about it, especially given the fact that, in pilot training, the airborne element of the course differs only slightly between countries. What part is played in pilot training by all the peripheral knowledge needed to fill out a degree course program? Is there a contribution to the quality of airline management when university-trained pilots start to take on managerial responsibility? At this stage in the development of aviation training, we are probably not in a position to answer these questions. However, the discussion does introduce a theme we will return to later in this chapter, and that is the development of 'expertise'. The view of training we take in this book is that our ultimate goal is to produce a workforce that can deploy expertise in order to overcome workplace problems. As training designers, we should be looking for ways to speed up the acquisition of this expertise.

By looking at differences in approaches to pilot training, we have seen that there is a concept of depth of knowledge that could be used to differentiate between systems in the way a particular syllabus item is treated. An alternative to 'time allocated to instruction' as a way of describing the extent of training to be provided is the concept of 'depth of knowledge indicators' found in the pre-Joint Aviation Authorities (JAA) UK pilot syllabus. Those for the Aircraft (Technical Group) subjects are given in Table 3.1.

Table 3.1
UK CAA Depth of Training Indicators

Depth Indicator	Definition
Code 1	Factual recall
Code 2	Comprehension and logical combination of separate items of factual recall.
Code 3	Logical combination of all relevant items of factual recall and a sound knowledge of the principles involved.
Code 4	A thorough knowledge of the subject and the principles involved, and their application to the problem.

Although each of the statements in the table can be challenged, they serve our immediate purpose in that they convey the idea that students are not required simply to receive and store information. There is some additional processing required before students can use the knowledge they have gained.

So far we have considered learning more as a function of the organisation within which it is conducted. The discussion has almost been a continuation of the previous Chapter in that national goals or the whims of the employment market have been used as the starting point. We now need to shift to the perspective to that of the individual within the learning system. We will now start to consider some differences between learners that will have an effect on the efficiency of the course you design.

Children v Adults

Probably the most obvious difference we need to take into consideration is that, whereas most of our formal education takes place in our childhood years, the trainees we have to deal with, as training designers, are adults.

It is well known, but often ignored, that adults and children go about learning in different ways. Whereas children are able to sit in class and have vast amounts on information thrown at them, adults are different. Adults need to be motivated to learn; they need to see the relevance of what they are learning and they need to be able to discover things for themselves. This 'adult' approach to learning is best described by Kolb's Learning Cycle,[5] summarised in Figure 3.1.

The idea behind the model is that, first of all, adults experience some event or aspect of their daily lives. They gather data about something. Next, they sit back and think about what is happening. They try to consider cause and effect. They think about the implications of what happened and try to spot any gaps in their understanding. Next, they try to generate some rules to explain what happened. They construct generalisations that can be applied to other, similar, situations. Finally, they set about testing those ideas in new situations. With increasing experience, our set of rules becomes more refined and we are more successful in our daily encounters. It is a cycle because we are repeatedly exposed to concrete experiences which, in turn, feed back into our learning.

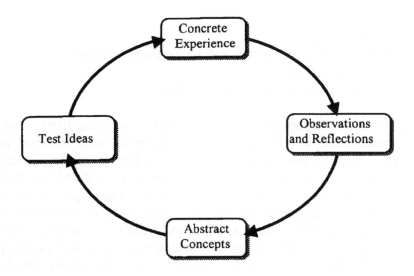

Figure 3.1
Kolb's Learning Cycle

We will discuss the implications of this model further in the next section on individual differences between learners. The idea we want to promote at this stage is that learning is an interactive process that requires learners to work with their subject matter in order to make sense of it. By 'subject matter' we mean everything from the principles of aerodynamics through to why a passenger has just poured their drink over the person seated in the row in front of them. Kolb's Learning Cycle reflects an informal mechanism by which we learn from life. We now need to look at some additional ways in which individuals in a class may differ.

Individual Differences between Learners

We once asked some students on a full-time commercial pilot course what they understood by the term 'learning'. Here are some of the things they said:

> 'learning something new or in more detail'
> 'learning how something works'
> 'looking for what you want to know'
> 'working hard - putting in effort'

'memorising'
'passing exams'
'learning something which will be useful for flying'
'gaining knowledge'
'knowing things I did not know before'

Some of these statements relate to activity -'working hard - putting in effort, memorising' - while others are concerned with the end result - 'learning something which will be useful for flying, passing exams'. A third distinction -'learning something new or in more detail' - typifies the view of learning as the accumulation of data. The depth of knowledge indicators mentioned earlier are similarly linked to the idea of the number of facts associated with a topic which must be fed to the student. All of the statements reflect a model of learning which involves effort on the part of the student and which results in the acquisition of information. Models that are based on this concept of 'storage' are concerned with the efficient placing of data in long-term memory and its subsequent retrieval. Although this is important, it is not really what we are trying to achieve in most vocational training systems. The knowledge imparted is only of use if it enables students to understand the world around them and, more importantly, to solve problems; hence the emphasis we place on learning as a mechanism in the development of expertise.

What the observations of these student pilots also reflect is the fact that we all bring different perspectives, qualities and characteristics to learning situations. The most significant differences between individuals are:

Aptitude
Personality and Learning Style
Attitude and Motivation
Information Processing (memory span, processing speed)
Prior Experience
Stress

Aptitude

Aptitude is an apparent preference or ability for a particular task or area of study. We all differ in our aptitude. Furthermore, ability in one area of expertise does not guarantee ability in all areas. We can all readily identify with the idea of someone who is good at science subjects but poor at the arts, and vice versa. However, there is a counter view that the skills needed to, say, learn a language are just as relevant to mastering calculus but the fault lies with educational systems which fail to teach learning skills in a manner which allows for transferability across areas of study. The fact remains, though, that individuals seem to differ in their aptitude for parts of the curriculum. For example, in many pilot Ground Training systems, Meteorology and Radio Navigation seem to be the most frequently failed exams, a pattern repeated across different nationalities. As subjects, they share a level of abstraction and deal with concepts that are hard to visualise. Technical subjects, such as aircraft structures, and practical subjects, such as flight planning, have better pass rates. Although there is no evidence to support the contention, it does seem that the pilot training syllabus requires different skills for some of the topics.

A more important aspect of aptitude, however, is its use as a criterion for selection for entry into training in the first place. It was not so long ago when 'an interest in mechanical things' was considered an indicator of suitability for employment as a pilot. Where a selection process has been set up, it is intended to identify aptitude and to select those individuals who stand a chance of succeeding in training. No selection system is perfect and attempts to increase the probability of correct selection also increase the probability of rejecting individuals who possess the aptitude we are looking for but who do not perform well at selection.

We should also bear in mind that past performance is no guarantee of future success. Many recruitment schemes require certain minimum standards of attainment in school. Unfortunately, there is no correlation between success at school and success in later life. Administratively, we can use such criteria - prior success in education - to reduce our pool of applicants to something that is manageable but we need to be wary when using such criteria to predict future performance. But some pilot training candidates are self-selecting. They have the money and the desire and it is up to the training system to help them to meet the standard. In these circumstances, aptitude is not a controllable variable.

Personality and Learning Style

Like aptitude, an individual's personality can influence performance in training although it is not something that can be easily changed over the lifetime of a training event. However, for our purposes, the implication of personality as a student variable is that some methods of instruction suit some people better than others. In a group of students, the chances are that we will need a mix of teaching methods if we are going to be equally as effective across the whole group. This idea has been developed into a concept of individual learning styles that reflect different preferences for particular types of learning activity. In the model developed by Honey and Mumford,[6] Activists revel in concrete experiences; they like to get their hands dirty. Theorists, on the other hand, are happiest trying to construct explanations. Reflectors like to watch others and learn from their mistakes while Pragmatists want to know the rules of the game. However, if our preferred approach to learning acts as a blockage, then we are probably not gaining maximum value from learning opportunities. We need to be aware of our preferred style and to take positive action to ensure that we exploit all 4 styles in the model. It should be stressed, at this stage, that Honey and Mumford make no claims for learning styles to be linked to dimensions of personality. They simply make the point that our style is the result of learned behaviour over time. That said, there does seem to be some fundamental relationship between style and behaviour to the extent that we might argue the case for styles to analogous with some dimensions of personality. For example, Activists and Reflectors may not show any difference in basic aptitude or achievement at the end of training. However, it does seem that Activists can get bored with mundane tasks more readily than Reflectors, in the same way that Extroverts seem to require greater levels of stimulation than do Introverts.

As an example of how learning style can influence students' reactions to training, we can consider a CRM course we ran for a group of new cabin crew. The course included two group exercises. One was a maze game that required groups to find their way back to safety. The second exercise was a simulation based on the aircraft passenger boarding process. In the first game, the group worked together to solve the problem while in the second game individuals took on roles outside their normal experience. The aim of the first game was to give experience of group decision-making and communication while the second exercise was designed to give insight to the roles of other team members during aircraft boarding (gate, pilots, ramp and so on). After the course we found significant differences between the way Activists and Reflectors viewed the two games. The maze game required no specific expertise and was highly rated by both groups. However, the boarding simulation was marked down by the Activists but rated highly by Reflectors. On subsequent questioning, we found that the Reflectors enjoyed the insight the game gave them into the work of other agencies who they rarely met and, in reality, could not even see at work during the boarding process. The Activists, on the other hand, became frustrated at their lack of understanding of the specific tasks they had to accomplish during the exercise, even though the game had been quite deliberately designed so that players did not need actual experience. We also found that Activists had difficulty seeing the big picture, which is partly what the game was all about, and this also added to their frustration. Accepting the earlier caveat about the relationship between learning style and personality, this example demonstrates that an enduring aspect of behaviour, the learning style of the individual, can have an effect on performance in training.

Attitude and Motivation

Attitudes are often considered to be pre-dispositions to react in a particular way in relation to some object in our environment. They are not as deep-rooted as personality traits and are more amenable to change in the short term. They are also fairly important in terms of student attainment. If we think back to the students' views of learning, 'learning something new or in more detail, learning how something works, working hard - putting in effort, passing exams' could all be considered representative of attitudes towards study as an activity. As such, inefficient study habits, triggered by an inappropriate attitude, can lead to under-achievement.

Attitudes to learning can be reflected in the particular learning strategy adopted by the individual. By 'learning strategy, we mean the way a student will process the material to be learnt. Three distinct approaches have been identified: shallow, deep and achieving Shallow processing is typified by rote learning while deep processing involves making an effort to understand and make sense of the course content. An achieving approach is motivated by the need to do better than fellow students. Moore and Telfer,[7] working with experienced commercial pilots, found evidence that the three approaches were apparent in the way pilots approached the learning task. When asked to rate their attitudes to learning, the pilots, on average, agreed with statements related to deep-orientated items, tended to agree with achieving items but disagreed with surface items.

The reference to an 'achieving' learning strategy introduced an element of motivation into the behaviour of the learner. An indication of a student's motivation can sometimes be found in the reasons given for starting a course. When asked why they wanted to be pilots, here is what some of the students questioned earlier said:

'I enjoy flying'
'Flying is interesting'
'It's a childhood ambition'
'I want to earn more money'
'My father was a navigator in the military'
'My father, uncle and brother are all pilots'
'I saw it advertised and it seems like a good job'

What these responses reveal is a dichotomy between intrinsic motivation - the appeal of the subject- and extrinsic factors - forces external to the course - such as earning more money. The first 3 statements in the list above reflect intrinsic motivation while the last 4 are good examples of extrinsic motivation.

Students who show intrinsic motivation tend to do better than their extrinsically motivated classmates. In a study of technician students on a distance learning course, those in the intrinsic motivation group achieved 6 grade A or B on their assignments as opposed to 1 for the extrinsic group. Conversely, the intrinsic group gained 2 grade Cs or D compared with 7 gained by the extrinsic group.

Intrinsic motivation usually leads to deep learning processes. Thus, because you find the subject interesting, you want to find out as much as possible about it. Extrinsic motivation leads to shallow processing. Quite often, ab-initio ground-school students are only concerned with learning enough to pass the exam and the same cry can be heard on type conversion courses for fairly advanced aircraft. Because the focus is on passing the exam, students simply try to learn responses by rote. In fact, an examination of Australian and Canadian flying instructors[8] found that the predominant style of instruction encouraged shallow approaches to learning on the part of students.

Whereas aptitude and personality are difficult to manipulate in the time scale of the average training course, attitudes clearly are. In particular, these linked ideas of extrinsic *vs* intrinsic motivation and shallow, deep and achieving approaches to learning give a pointer to the need for training designers to build in guidance to students to help develop extrinsic, deep processing strategies. We will discuss this aspect in more detail in Chapter 6.

Information Processing

Another feature of the student population we need to consider is that of individual differences between students. Essentially, we all show variation in such mental and physiological characteristics as memory span and speed of mental processing. Students will also differ in the extent to which they can see important messages hidden in background noise. Known as Field Dependency (FD), students who demonstrate high FD need guidance in extracting important messages whereas students who have low FD are able to identify key points more readily. Here are 2 versions of the same test question to illustrate the point:

a. $2 + 2 = ?$
b. I was walking down the road with a friend. We were going to a party. On the way we met 2 people going to the same party so we all agreed to go together. How many of us arrived at the party?

In theory, the low FD student should do equally as well with both styles of question whereas the high FD student will find questions framed like Version b more difficult than Version a.

We can also identify differences between students in terms of those who take an holistic approach, that is, they look at the big picture, as opposed to those who work sequentially through the subject matter from start to finish. Others work with visual representations in preference to verbal descriptions. When we consider the needs of learners, we need to take these differences in style into account.

Prior Experience

It may be a statement of the obvious but students will all bring different levels of prior experience to the course. We will talk more of prior experience in Chapter 6 and in the section on expertise later in this chapter. We need to be mindful of prior experience because it establishes a context and makes the acquisition of new information easier. Unfortunately, one of the differences between learners that we also need to note is that they are not all equally capable of recalling prior experience. Learners differ in their ability to establish the relevance of events and experiences. If they do not see the relevance of an event, they will not bring the information into working memory. As training designers, this will sometimes be our job!

Stress

The final individual difference we need to be aware of is stress. We all differ in our tolerance of stress but we need to be aware of the fact that stress is a significant obstacle to learning. It is one of the few individual differences between students that has been found to have a clear correlation with performance in training.

So far we have looked at learning in terms of the system of delivery and also from the perspective of individual differences. We have seen that learning is greatly affected by things outside the learner and also by some fundamental characteristics of learners themselves. We now want to look at another aspect of learners, their method of interacting with the subject to be learnt.

Learning as Activity

Having considered how our learners differ from one another, we will now move on to consider what it is learners do in order to accomplish our goal of actually learning something. In Figure 3.2 we illustrate the problem we have to deal with. On the one hand, we have some subject matter that relates to the activity or topic being covered on our course. On the other hand, we have a set of hidden processes which take place within the student. As trainers, we sit between these two and offer a set of activities which will allow learning to occur. The next section will concern itself with the centre circle on our diagram.

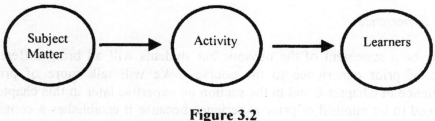

Figure 3.2
A Model of Learning

Tactics of Learning

Effective learners in formal settings use a variety of tactics to promote their understanding of the subject matter. A study of ab-initio pilots[9] identified

60 techniques used during class work ranging from working through given examples, through preparing summaries, inventing mnemonics, seeking further information to preparing lists. The techniques broke down into 7 key groups:

- Techniques involving practice.
- Identifying key ideas.
- Relating information to other contexts.
- Techniques to aid remembering.
- Filling in background.
- Projecting and predicting.
- Grouping or clustering.

Different types of learning activity, such as writing an essay or dealing with a case study, require different techniques and students differ in the effectiveness with which they use the techniques. However, the study reinforces the view that learning requires effort on the part of those involved. Whereas Kolb's model can be seen as an informal mechanism, the study techniques are clearly formal tools used during learning. In both cases, students differ in the extent to which they are effective in the process. What is important is that we can see learning as something to do with the way people interact with the subject matter. The next thing we will consider is what goes on inside the student while learning is taking place.

Types of Learning

The views of those student pilots cited earlier who said that learning is about 'something which will be useful for flying' or 'learning how something works' reveal that the end result of training needs to be something of use to us. The storage of data in isolation, probably something that most of us experienced at school, serves little purpose. It is only when data is rearranged in a meaningful framework that real learning take place.

Most of the individual differences between learners discussed earlier, their attitudes and motivation can be measured to some degree. For the rest of this chapter we will be dealing with aspects of learning which, to a large degree, are hidden; they are conjectures. We need to consider what is going on inside the student's head while learning is taking place.

To follow on from the idea of converting data into knowledge through association, it seems that learning is a process that requires individuals to learn ways of interpreting and predicting the world around them. This view implies that learning is a social process and requires students to interact with models of the world so as to better-structure their own internal representation of the world within which they operate. Unfortunately, though, despite our training methods being all too visible, their impact on students' internal learning processes is invisible. We can only deduce what is happening from such measures as exam scores or time to proficiency. In short: observable performance.

To fully understand what is going on during the act of learning, we need to distinguish between the process and outcome. By process, we mean the intervening stage between what the student is being subjected to during training and what ends up getting stored in long-term memory. By outcome we mean the different ways knowledge is coded in memory.

An early attempt to identify intervening processes was Bloom's Taxonomy of Learning[10] that lists the following stages in developing understanding:

> Memory Recall
> Understanding
> Application
> Analysis
> Synthesis

Memory recall is, quite simply, bringing information out of storage. Understanding requires us to be able to explain an item by either changing its form, as we might do to solve an algebraic equation, to describe possible trends. Application requires us to be able to take an abstract form and apply it in a concrete situation. Thus, in meteorology we might be asked to use Boyle's Law to describe some aspect of the behaviour of the atmosphere. Analysis involves looking for the relationships and principles within some information. For example, when we examine air safety incident reports we will see that some events occurred while the pilot was distracted. By applying our skills of analysis we will see that the common link is a failure to share attention adequately. Finally, Synthesis involves reorganising information into new structures. So, from our knowledge of the aircraft hydraulic system, we can devise a solution to a malfunction that is not contained in the Abnormal or Emergency check-lists.

Taxonomies of learning, such as Bloom's, describe tasks we can set the student which will reveal something of what is going on when learning occurs. We observe learning through inferences based on student actions.

Another theory identifies 4 key learning processes: associative learning, procedural learning, inductive reasoning and metacognition. Association is the process by which relationships are built up between ideas. Procedural learning is the acquisition of sets of rules which explain why things happen the way they do. Inductive reasoning occurs when new rules and principles are discovered as a result of processing other information. Metacognition is a form of top-level management, which keeps track of how our learning is progressing. It allows us to check how well we understand a subject and devise ways of improving our knowledge. We can further break down these types of learning to identify the conditions that will support learning;

Association
    Rote - or learning by memorisation.
    Didactic - or learning from instruction.
Procedural
    Practice - or learning by doing.
Induction
    Analogy - or learning from something similar.
    Examples - or learning from more of the same.
    Discovery - or learning without any help.

Metacognition has been left off of the list as it is an over-arching concept that allows us to manage all of the learning processes listed here. In part, a reflection of our metacognitive skills would be the effectiveness with which we used the tactics of learning described earlier in this chapter.

The 4 learning processes have been described as intervening stages between the subject matter under study and the storage of information in memory. The way information is stored, the third circle in Figure 3.2, will be considered next.

Internal Learning Processes

We need to remind ourselves that, despite advances in magnetic resonance imaging, which is allowing researchers to track brain activity in real time, we still do not know exactly how information is stored in the brain. Thus, our conclusions about memory storage need to be based on the outcomes of learning; the observable elements that arise as a result of learning. It seems that internal processes transform data into either subject knowledge or regulatory activities that then enable us to convert that subject knowledge into actions.   One view is that knowledge can be divided into two categories: Declarative, or the things we say and Procedural, or the things we do.   More specifically, declarative knowledge comprises propositions, schema and mental models.   Propositions can be considered the building blocks of knowledge.   They are individual pieces of information which can be grouped together to form concepts and which can be qualified by adding facts.   We can pull together collections of propositions about a particular situation and, now, we have a schema.   Schemata help us to make sense of the world.   They act as a filter through which we receive new information. By adding rules, which explain behaviour, to our schema we generate mental models.   Mental models differ from schema in terms of their dynamic properties that enable us to predict outcomes and to run what-if scenarios in order to decide on possible courses of action.

Procedural knowledge comprises the rules and skills that we use to enable us to act within the world.   Rules are simple sets of IF-THEN statements that allow us to work out dynamic relationships between propositions.   More complex sets of rules build into skills which, after continual practice can become automatic, such that they can be completed with little mental effort on our part.   Physical skills can be included in this scheme in that they are rules translated into action.

Learning, then, implies the building of internal knowledge structures, establishing inter-relationships between parts of those structures and then fine tuning as a result of repeated use.   The process is considered to be generative and constructive. Generative in the sense that students generate meaning out of material presented to them and constructive in that students can construct skills and knowledge through interactions with their surroundings. By generating meaning, students can relate new information to their existing internal structures; which, in turn, are the result of prior experience discussed earlier.   By testing knowledge in the environment, students re-organise their mental structures. It is also clear that this process can result in false knowledge, or errors. For example, a faulty schemata could cause a new piece of information to be misinterpreted, leading to an incorrect view of why an event occurred.   To illustrate the point, an experienced Captain learning to fly a new type of medium-sized passenger aircraft, having finished and passed ground school, was now completing the airborne exercises needed for final qualification.   After one particular flight, the Captain reported a malfunction with the hydraulic system. On questioning, the Captain reported that the aircraft's elevators had seized because of the failure of the hydraulic system supplying the aircraft control surfaces. In fact, the aircraft concerned had mechanically, not hydraulically, -controlled elevators. The Captains schema for aircraft controls was faulty resulting in his generating a false interpretation of an event. In fact, the aircraft had been sitting in the rain, had passed through cloud while climbing to height and water had frozen the hinges of the control surfaces.

Developing Expertise

We have considered learning in terms of characteristics of the training organisation and in terms of attributes of learners. We have also looked at the way learners interact with the world, the intervening process by which information from the world is manipulated and, then, how knowledge is stored. However, the story of the ill-informed Captain throws up another aspect of training. We want people to solve problems. We want them to be able to keep the system functioning whilst possibly operating beyond the limits of their experience. In short, we want them to be experts.

It is often said that 'you cannot teach experience', and that is true. However, you can short-circuit a lot of the time wasted in trying to gain experience if you have a clear idea of what 'experience' means. Our ultimate goal as training designers must be to accelerate the acquisition of expertise, which is why we need to better understand just what it is we mean by 'expert'.

It seems that experts stand in relation to their field of expertise in a way that differs from novices. It appears that they can anticipate events, solve problems faster and maintain a level of skilled performance with less effort. Underlying this idea of expertise is the assumption that the knowledge domain - the particular field of interest - exists as an entity and that Experts have greater familiarity with its structure than do novices. Don Norman, author of 'The Psychology of Everyday Things',[11] suggests that we can identify the target system, or subject-matter, the conceptual model of that system and the users' internalised version of that model. We saw earlier that one suggestion for how we learn is the idea that we construct models of events, or objects, in which ideas are linked together to form functional relationships; schema and mental models. We can use our mental models to understand what is happening and to anticipate what is going to happen next. In this case, the target system is an absolute version of the real world. The conceptual model is a form of description that converts the absolute version into structures that can be communicated and stored as mental artifacts. The user's version is a representation of the conceptual model after their individual learning processes have mangled it.

From this description, we can perhaps consider expertise to be some state where the target system, conceptual and model and user's model all aligned as closely as possible. Unfortunately, there is a problem with this approach. We once did a Training Needs Analysis (TNA) for a course which was intended to teach experienced pilots the technical aspects of a new aircraft type. As part of the process we sat down to apply some of the techniques described in the next chapter. We were working with a pilot and a maintenance engineer. The engineer taught the Digital Flight Guidance System (DFGS) in the classroom to other engineers and to pilots. The pilot flew the aircraft and taught other pilots how to fly it. Neither of these two experts in their field could agree on what the course should contain. The engineer's view of the course content included some fundamental aspects of the design of automatic flight control systems. The pilot's view was more rooted in the daily operation of the system. If you tried to get this pair to list everything which was essential to know about the DFGS, you would get two very different lists.

In fact, they both have different views of what we called earlier the target system. We can see differences like this in other areas of training. For example, Human Factors, as a topic in a former-Soviet aviation training system, is an extension of Ergonomics whereas in a UK curriculum it is seen as a development of Aviation Medicine. The subject addresses the same problems but come from very different directions. Possibly, then, knowledge does not exist in any absolute sense. In which case, how valid is the concept of a target system that can be codified, via a conceptual model, in terms of a mental model?

Another difficulty with the target system/user's model proximity view of expertise is that, in completing a task, we do not simply restate the structure of our mental models. Instead, we draw inferences from our internal representation of the world in order to solve problems. So, another view of mental models is that, in fact, what we try to do is to gather a collection of 'problem spaces'. A 'problem space' is a mental representation of a collection of possible states of the world, together with rules for moving between states. If you accept that most of working life is about fixing things that go wrong, then a part of experience (and, by association, expertise) is the fact that 'we have seen it all before'. As we gather experiences, we see the types of problem that can arise and we witness solutions being applied. Of course, no one person can experience every single possible problem - there is always a new way being invented to mess things up!

We can explore this idea by considering someone working at an airport check-in desk. We know that passengers come in all shapes and sizes, and they all have different ways of dealing with the process of being checked-in. Most fall into the category of 'easy to deal with'. Others can be awkward and even downright abusive. Through experience, Passenger Handling agents have come to recognise the tell-tale signs of a difficult customer. For example, if single male in a business suit and carrying a briefcase approaches the check-in desk with a broad, friendly smile on his face, then expect a request for a free seat upgrade. As a result Passenger Handling agents have developed a set of techniques to use to reduce stress induced by these people and keep the process moving along smoothly. Their problem space is the complete range of encounters they could have with passengers at check-in. Their stress reduction techniques are just some of the rules they use to move from one type of problem to either a satisfactory solution or to some intermediate version of the problem - such as calling the supervisor and letting them deal with it. Quite often we call these rules heuristics or 'rules-of-thumb'.

Now, we need to refine this approach slightly. In real life, we do not passively wait until a problem occurs before we take action. In almost all that we do, we have a goal in mind; we are goal-directed. For example, at check-in, the usual goal is to process the customer quickly and efficiently, keep a smile on our face and, at the same time, make sure that the passengers leave as much carry-on baggage at check-in as possible. We may have some other goals that come into play when, say, a passenger does try to sweet-talk an upgrade or another who refuses to be parted from their clearly overweight hand baggage. A part of expertise is in selecting the right goal from all of those available to us at the time.

Expertise can be seen as the possession of a set of extensive and accurate schema, together with a set of rules for extrapolating into new areas. From the few studies which have been conducted on the differences between experts and novices, it is not clear that experts necessarily possess more facts that novices. Rather, it is the way information is organised which makes the difference. Expertise is also characterised by an apparent effortless application of the stored schema. In part, this is because experts seem better able to break down problems to their component parts before looking for solutions. Novices seem to work in reverse; they apply what they know to create a solution rather than, first, trying to find what is the problem. Experts seem to work in an effortless manner because they have reached a stage where their working practices have become automatic and, for this reason, experts often have difficulty explaining exactly how they go about solving problems. Part of training design is to make clear the schema required to understand and control activity, and that is what we will cover in the next chapter.

Conclusion

As part of your training design activity, having identified any strategic, organisational issues, you now need to consider some problems posed at the level of the individual trainee. For example, we should not forget that adults are different to schoolchildren in their approach to learning. We need to think what effect individual approaches to learning will have on our project. We said that one of the difficulties with learning is that we cannot see what goes on inside a person's head; we can only deduce what is happening from indirect but observable measures. Lest we get carried away, we should add at this point that attempts to prove the need for training to be adapted to individual styles has delivered mixed results. All we want to do at this stage is to identify the need to consider the implications of how learning occurs. We will look at the activities of learning in a later chapter. Your job is to combine a set of activities that will provide the opportunity for optimal learning to occur.

Case Study

How many of you have travelled on an aircraft and tried to make sense of the safety instruction leaflet? In particular, have you ever looked at the diagrams that try to explain how to open the emergency doors and then asked yourself if you could ever do it for real? In some airports, they have tried making training doors available in the departure lounge for any interested passenger to try out. Rather like fighter ejection seats and car airbags, devices which must work first time every time, opening the emergency exit is something which people also need to get right first time but without the chance to rehearse.

Imagine that you were going to design a training course to teach passengers how to open the emergency exits. How would you need to shape your course in the light of our discussion about learning theory? Consider the change in behaviour - what is the task to be learnt? Consider student characteristics. What organisational factors will need to be recognised? How will these affect learning process?

4 Analysing the Requirement

Introduction

In the last chapter we spent some considerable time reviewing what we meant by learning. We needed to labour the point because, after all, as training designers we are setting out to enable others to learn. The next stage in the process is to identify what it is we want people to know and do at the end of our course. Nor should we forget the requirement to include some insight into how to do the job in an expert fashion. In this chapter we start the process of training design proper. We will be covering 3 main areas in terms of analysing the job of work: what to look for, how to find it and, finally, how to describe it. At the end of this chapter you will have:

- Considered the component parts of job performance.
- Considered a variety of techniques for identifying the parts of job performance.
- Considered a range of ways of describing your course syllabus.

What are we Looking For?

From our discussion of learning and, in particular, expertise, we can see that we need to identify 2 groups of activities. First, we need to identify what it is people physically do as part of the job for which they are being trained. Then, we need to identify the controlling activities people use in order to be expert at doing their jobs. We can broadly categorise these things as 'action' and 'control-of-action'.

Under the heading of 'action' we are interested in the things we do in order to get the job done. These include the physical skills, like manually controlling aircraft, as well as procedural tasks, like filling in documentation. We can start to draw up a task list. We can include details of different agencies people liaise with during the work cycle and also the various, tools, equipment and documents used in completing the task. We should also include all of those things we do to configure the workplace in order to achieve the task goals that apply at a particular time. We can describe the inputs required at various stages, including their source, and we can also list the outputs from the work process and their destination. We give examples of task listings later in this chapter.

Under the heading of 'control-of-action' we can include all of those things we do in order to move from one problem-state to another in order to achieve our goal. Control-of-action is a term which describes all of the mental processes involved in identifying what the problem is, seeking information, formulating plans, monitoring progress and so on. It includes all of the routines we run, automatically, as we go about our business. And it is important to remember that a large amount of what we do is automatic behaviour. The evolutionary process has done wonders in ensuring that, as far as possible, we can get through the day with little conscious effort. More important, though, is the need to identify how people do these things. So, what are the signs that a task is starting to go wrong? How do people re-allocate priorities? At this point we start to look for the 'rules-of-thumb' people use in order to keep things on track.

The categories are not watertight. For example, the 'action' of a baggage handler loading suitcases is intimately bound up with the 'control-of-action' processes such as guarding against workplace injuries, concern that aircraft and baggage are not damaged, efficient placing of the bag so that it can be stowed properly and so on. But the distinction serves our purpose in underlining the difference between observable actions and unobservable mental activity.

We can elaborate on the list of areas to investigate by considering a theoretical model of a job. In Figure 4.1 we have represented the characteristics of a job.

The idea behind the model is that work performance, the Outcomes in the diagram, is a function of some core aspects of the job and certain psychological states that prevail within the worker. As part of our analysis, as well as identifying the component parts of the job, we can explore some of these psychological states, at least to the extent that they are reflected in workers' attitudes towards their job.

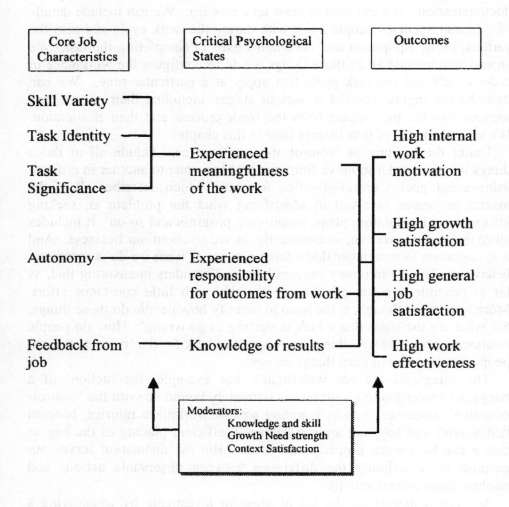

Figure 4.1
Job Characteristics Model[12]

We need to add a word of warning at this stage. The simple description of the task is probably better known as a Task Analysis.[13] We need an acceptable task description as the basis for our work but, at some stage, we also need to identify what effective operators need to know in order to perform in a competent manner. More important, we need to start thinking about what training will be required in order to become proficient at the task. Figure 4.2 illustrates what we mean. We make the point here because, all too often, many TNAs seem to stop at the level of task description.

Task Analysis   Training Needs
Analysis

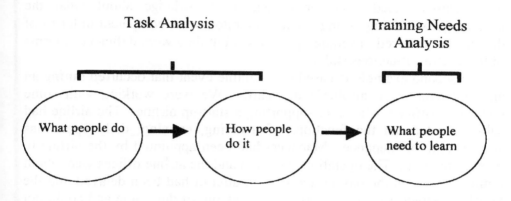

Figure 4.2
TA vs TNA

To make this distinction a little more clear, we want to look at a couple of examples. At the same time, we want to reinforce the importance of getting past the surface features of the job in order to explore fully how the job is managed at a cognitive level. We believe that exploring expertise is a neglected area of task analysis, probably because of the difficulty involved in getting to the heart of expert performance, but in such a risky endeavour as aviation, it is here that the greatest payoff will be found. The first example concerns a UK airline we were working with which noticed that some of its pilots were not using the correct technique for managing the Flight Management System (FMS) when the aircraft was in the descent. On at least one occasion, the aircraft ended up sufficiently high on final approach that the pilots had to go around for a second attempt to land, all of which costs time and money. All of the pilots had successfully completed the aircraft conversion course and many of them had several years experience of flying this particular type. There had been no changes in procedures or equipment recently. It transpires that, although all of the pilots knew how to set up the FMS for the descent, they did not all fully understand how the FMS managed the descent profile. Therefore, whenever the pilots had some options for configuring the FMS, any variations from the norm could result in awkward and dangerous situations developing. Part of the problem was that the training course was designed around a description of the steps needed to set up the FMS found in the manufacturers' manual; a task analysis. The manual failed to consider what learners need in order to acquire knowledge about what the technology was doing during flight. The pilots were competent in terms of the actions involved in setting up the FMS but they were deficient in terms of the control-of-actions skills.

The second example is based on a routine event that occurred during an airline operations controller's day shift. We were working with some operations officers who were supporting a start-up airline. The airline had sub-contracted out the functions of rostering, crewing, daily operations support and maintenance. Managers had been appointed by the airline to oversee the task. The operations function and the airline offices were about a mile apart. On the day in question an aircraft had been delayed and the knock-on effect was that the crew was now out of duty time and could not make the return flight back to home base. The duty operations controller, in order to keep the schedule vaguely on track, called out a standby crew and sent them by taxi the 200 miles to the airport where the aircraft was sitting with its load of passengers. The bill for the taxi fare was something in the region of £300. The airline operations manager complained to the sub-contractor that the solution worked by the duty operations controller was unacceptable.

In terms of task analysis, we are interested in what the controller did, why he did it and the effectiveness of his actions. In the absence of any other instructions, the controller saw a problem - an aircraft unable to complete a scheduled flight - and implemented a solution - dispatched a second crew by road. The controller had evaluated the alternatives and decided that this was the most expedient solution. So what was the cause of the airline managers complaint? In short, the £300 taxi bill. The operations controller had considered the need for a minimum delay for the passengers as part of his planning. However, the airline manager's alternative was to re-route another aircraft which had capacity and whose crew had sufficient duty hours to accomplish the task. The delays caused to the passengers on the original route and those to be collected by the diverted aircraft did not feature in the planning. Control of cash budgets was more important at this stage of the airline's growth; passenger satisfaction was lower down the list of priorities. In this example, the first problem we encounter is a poorly defined task, in that responsibilities are obscure, compounded by inadequate information concerning priorities to be applied when arriving at a problem solution. The situation was made all the more difficult by having the job shared between the sub-contractor's Ops Officer and the airline's Ops Manager. The story neatly illustrates some of the problems of identifying expertise, in this case that of the operations controller in fixing real problems. The effectiveness of an individual in the workplace can depend upon your point of view, hence our discussion of stakeholder analysis in Chapter 2. The controller's solution could never have been considered successful because he was not made aware of all of the constraints on possible actions. As part of our analysis we need to explore different perceptions of the job and grades of success. Some questions which can be used to form the basis of a task analysis are given in Table 4.1.

Table 4.1
Task Analysis - A Basic Checklist

- What activities are involved in doing the job?
- Do you use physical skills?
- Do you apply complex procedures?
- What is the task frequency?
- What is the task difficulty?
- What is the task criticality?
- What objects do you use as part of the job?
- Do you analyse information or interpret documents, tables or charts?
- Who do you communicate with as part of the job?
- What information do you need from others?
- Who do you pass information to?
- What is de-motivating about the job?
- What knowledge do you have of the results of the job?
- Are you required to check the progress of events?
- Do you discriminate between degrees of performance?
- Do you allocate priorities or plan work?
- Do you look ahead and anticipate events?
- Do you diagnose problems?
- Do you create new solutions to solve problems?
- Are you required to adapt actions to new situations?

How to Find It - Analysing the Task

We need, then, a set of tools which will allow us to identify the actions associated with doing a job and also the control structures which regulate how we complete those actions. Immediately, we hit a problem. We can see the 'actions' bit but the 'control-of-actions' is invisible, but more of this later. The methods available for the analysis task fall into 2 categories:

> Direct - watching people at work, watching recordings of people at work or learn to do the job ourselves.

> Indirect - interview people about how they or others do the job; looking at outputs from activities (records, etc); reading manuals, instructions and orders; look at incident reports and customer complaints.

Direct Methods

We need to offer a word of warning at this stage. Observation of performance is probably the most widespread technique used in training design. In fact, often training is run by people already experienced in the job and the course is based on their prior experience; the observation has been done at a distance. This tendency has led to a weakness in course design; training is often confused with task rehearsal. Thus, the instructor describes the job and, once they have finished, the student will have the chance to practice and, hey presto, training has occurred! This approach has worked well for years as is demonstrated by the use of flight simulators for pilot training. By rehearsing in the simulator the procedures to be flown in the aircraft, there is an assumption that we are building skills. However, returning to our model of expertise, can we be certain that the activities to which the student is being exposed will be translated into effective knowledge structures? Can we be certain that acceptable performance in the simulator is the output from an accurate (or, more properly, effective) mental model or was the pilot just lucky on the day? One research project which looked at training pilots in the skills of instrument flying found that what instructors taught and how they themselves scanned the instruments were, in fact, remarkably different. The point being that our purpose in conducting a task analysis is to identify what people do, not what they or others think they do.

The direct observation methods include shadowing a trained worker as they go through a task or learning to do the job ourselves. Where observation of others is involved, we may want to interrupt the process to get the person being observed to explain what they are trying to accomplish. The questions in Table 4.1 will serve as the basis for a framework of analysis.

It should be immediately apparent that direct observation can be time-consuming and also very intrusive; people generally do not like being watched while they work. Therefore, we need to allow time to establish the right relationship with the people being observed. Another problem is that we cannot always guarantee to witness the full range of activity involved in the job, particularly those problem areas where people may be required to demonstrate expertise. We are not always given access to exemplars in terms of performance. We once spent a day shadowing an operator only to be told later by a manager that the views of the individual should not be taken seriously, as the person was not considered to be a good worker. At the other extreme, though, it is not unusual for an organisation to protect the analyst from poor performers in an attempt to create a good impression, or to limit damage. On one particular occasion we specifically asked to be given access to some personnel who, in the opinion of management, were considered to be at the lower end of the performance scale. The request was denied.

We can use second-hand methods, such as videotaping people at work or by getting operators to keep diaries of the things they do. Unfortunately, much information is lost this way because points of detail may not be recorded and, of course, the performer is not always available to explain what they were doing.

The output from this sort of activity is similar to that required by Human Resources Departments to define job specifications, agree job adverts and to establish pay scales. There is a risk, therefore, that we can end up with a highly-detailed description of the job but little useful information upon which to base the course design process. In part, this is why we issued the caution earlier about the confusion between a Task Analysis and a Training Needs Analysis. Although the techniques appear the same, the data will be put to very different uses, and this should be borne in mind at all times.

We said that observation can be tedious. One of the reasons for this is that it can sometimes be difficult to set the lower bounds of the analysis; it is easy to get sucked into trying to describe the minutiae of the job. As a rule of thumb, if something is either so simple that a reasonable person could pick it up without specific instruction or so unimportant that it doesn't matter if we get it wrong, then it probably should not be included it in the analysis. At all times, we need to remember that we are using the description of the task as a starting point for the identification of the types of learning required to be proficient. If our analysis does not allow us to reach that goal, then we have been wasting our time.

Indirect Methods

Manuals

Indirect methods of analysis are those techniques which can be used away from the job being analysed. They involve examining any one or any thing that can give us information about how people do their jobs, the level of performance achieved and any problems they encounter. Again, there are some drawbacks to watch out for.   First, all documentary evidence (manuals, procedures, etc) is written by humans and so must be treated with suspicion. Quite often, the way the manual describes the job actually masks some of the relationships that exist in the real world.   Aircraft manuals are invariably written by engineers whose views about what students might need to know about the system differ from those of pilots, as we saw in Chapter 3 when we looked at the problem of the DFGS course.

Diagrams contained in manuals are often drawn for convenience rather than as an attempt to accurately portray conceptual or functional relationships. We once worked on a project to develop computer-based training to teach pilots about aircraft systems. We were asked to use the systems diagrams in the manufacturer's manuals whenever we wanted to portray a system on screen. During the prototyping of the hydraulic system lesson one of the maintenance engineers pointed out that not only was our on-screen portrayal of the system incorrect but so was the manual. Because the hydraulic system was relatively simple and most people knew how it worked, no emphasis had been placed on updating the manuals.

Procedures are often there to constrain performance in the real world; they provide a baseline framework within which tasks are to be conducted. Within the framework, operators are free to use a range of techniques to get the job done. Thus, published procedures alone do not necessarily reflect the richness of techniques used in the workplace.

Company Reporting Schemes

Incident reporting systems, where they exist, are invariably rather skewed samples of performance and can not be considered fully representative of how the workforce is performing. Quite often, the very issues we are interested in from a training perspective do not get reported. Most schemes still tend to focus on technical malfunctions as opposed to crew inefficiencies and the detail contained in the report is usually inadequate. The rise of Human Factors reporting and the development of sophisticated reporting formats are generating better material but this area is still very much in its infancy as far as training design is concerned. However, it is still a source of information, as are customer complaints, Health and Safety occurrence reports, the personnel department's sickness records, rates of turnover of personnel and so on. This group of information sources can give pointers to where we need to look in more detail. They are also useful for identifying training needs at the organisational level.

These documentary sources of information can provide clues that can then be followed up with interviews.

Interviews

We use interviews to confirm that our task analysis is complete and that we understand why things happen the way they do; we are trying to fill in the gaps around the performance as observed. We are also trying to get a feeling for the problems that students have in acquiring expertise. All too often, asking someone about a task or job will elicit a stream of things that other people are doing wrong. Interested parties all have an axe to grind and, so, to rely on the Fleet Manager's view of what his pilots need to be taught can result in a rather unbalanced view of the problem. Usually, you have to allow managers to have their chance to express their opinions before they can get round to your immediate task. Often, what they think are problems will be a perception of something happening in an unsatisfactory manner but with little or no evidence to support their contentions. General expressions of feeling or belief are not really helpful so whenever a problem area is identified, get the interviewee to give specific examples using the techniques described below. However, notwithstanding any concerns we may have about the validity of management views, where training has already been running for a while, instructors' views about where students have difficulty and where most failures arise can give valuable clues about shortfalls in instruction.

Most of the techniques described in the next section are designed to clarify issues around how jobs are done and to make concrete examples of good performance. We once did an exercise to identify what 'good aircraft captaincy' meant. Many of the people we asked replied 'I can't explain it but I know it when I see it'. Clearly, such responses are of little use when designing a Captaincy Course but, in fairness, most of us go through life adapting to and coping with situations but without ever stopping to analyse in explicit terms what is going on around us; we just do it.

Questionnaires

A final technique we need to consider, but one which is fraught with difficulty, is the use of questionnaires. Notoriously difficult to prepare and well known for their low response rates, they can often be a waste of time if not constructed properly. We can use questionnaires to determine existing levels of skill and knowledge prior to introducing training or to identify attitudes in the workforce. We can also use them to provide some quantitative feel for what has already emerged from interviews. In this case, we would use the output from the interviews or direct observation methods to construct a schedule or task listing. We can then use a questionnaire format to obtain information about how often people complete the task or how difficult they find it to be.

Structuring Interviews

Critical Incident Technique

Interviews are an important tool in conducting a TNA and we need some ways of bringing order to our investigation. The first approach we will look at is known as the Critical Incident technique. First, allow the candidate to identify broad areas of the job which either cause problems or could be considered indicators of skilled performance; basically, if you can cope with this problem then you can cope with anything. Get candidates to identify either an extremely good or extremely bad example of the activity under investigation. Get them to explain what happened, why it happened, what people did about it and how successful they were. Ask why, in their view, the incident was important. As a basic rule, an interview will prove more valuable if, for each event, we can get the interviewee to describe situations, task, actions and results. By this we mean what was the background to the event, what was supposed to happen, what actually happened and what was the outcome. We are trying to identify examples of good and bad practice in order to establish the standard of performance expected and the skills to be employed in the under investigation. Try to get a range of examples relating to the same aspect of the job. It is often easier to identify bad performance than good performance but you should try to get exemplars of perfection as well as disasters.

At the end of this exercise you should have a set of examples, which is always useful, but more important, you should have a set of explanations. We have laboured the need to get at expertise in analysing training needs and it is in the explanation of the event that the interviewee will reveal the components of expertise. The explanations should reveal associations between ideas, rules of thumb which are applied in assessing situations, techniques of problem-solving and so on. In many ways, the incident is only a vehicle for examining these other aspects of performance.

Repertory Grid Technique

Another technique that can be powerful under the right conditions is the Repertory Grid. The full procedure is time-consuming and can be difficult to use. However, the shorthand version, so to speak, requires the interviewee to identify a range of people whose work-place performance they are familiar with. They need to write some form of identification mark for each of the people on a separate piece of card. It is important to stress that the interviewer is not interested in who these people may be; we will simply be using them for comparison. You will need a range of performers, some good and some not so good. Shuffle the cards and extract three. Place two together on one side and put the third off to the other side. Ask the candidate to think of the three people represented by the cards and get them to try to think of a quality or characteristic, in relation to the task being investigated, which the pair share and which separates them from the singleton card. Once the distinguishing feature has been identified, ask for clarification and, if possible, an example of the behaviour in practice.

From our experience we have found that the comparison task can cause confusion. For example, because the cards are selected at random, it is conceivable that we will have a good and bad performer as the pair with another good performer as the singleton. The tendency is to compare good with bad and so the interviewee will mentally re-combine the selection in order to isolate the poor performer. You need to be on your guard to ensure that the method is applied as intended. It may well be that the interviewee cannot identify any distinguishing characteristic shared by the pair. If so, then simply reshuffle and try again. Keep working through the pack of cards, pulling out, at random, three cards at a time, until no further information is obtained.

A shortened method we have found to be useful involves getting the interviewee to identify a range of people, between six and ten, and to write their names or some other identifier on bits of card. Shuffle the cards and then draw out pairs. Get the interviewee to compare the two people named on the cards in terms of the aspect of the job under consideration. For example, you may say that you are interested in who would make a good senior cabin crew member. The interviewee draws up a list of senior cabin crew whose performance is known to them. As each pair is drawn out, we ask the interviewee to identify differences between the two individuals in terms of how they do the job. By repeating the pair-wise comparisons, we can draw up a list of effective and less-effective behaviours which can then form the basis of a training course.

The Repertory Grid technique has its roots in personality research and can often seem rather bewildering on first encounter. However, as a way of getting people to put into words what were formerly just vague ideas, it can prove to be very useful. At the end of the exercise you should have a list of statements related to the topic of investigation. We could get the interviewee to rank-order the list once the task is complete. This would give us an idea of significance. We can introduce the Critical Incident approach and get the interviewee to give an example from the workplace of each of the items on their list. The Grid technique is useful in generating rich descriptions of performance but does require a little practice.

Card Sorts

A simpler technique which can also be used to good effect is the Card Sort. You can use it as a preliminary to other techniques. For example, if some prior information about the task is available, write the individual elements on separate cards and get candidates to sort them into priority order. Then, using the cards in sequence, get candidates to identify Critical Incidents for the top five cards. From a syllabus of instruction or as a result of output from the Repertory Grid exercise, you can write ideas and concepts relating to an area of activity on separate cards and get candidates to identify relationships, degrees of importance or significance.

We can use a card sort approach to reveal the structure of expertise. First collect together a broad range of concepts relating to a job. Write one concept, idea, action or whatever on each card. Get the candidate to sort the cards into piles of related ideas. Once the task is complete, get the candidate to explain why groups of cards are related. By this method we can start to get at the conceptual structure being applied.

At the end of a card sort exercise you should have a set of related ideas, some possible labels for components of expertise, some start points for further enquiry. The card sort is a useful technique for bringing order to apparent chaos as well as useful way of opening up new lines of enquiry.

Focus Groups

So far we have talked about methods which are best used in a one-to-one situation. That said, they could all be used with groups. As part of our investigation process we sometimes need to get groups of people together to seek their views. The group involved in the process needs to chosen with care and it is probably worth considering running several sessions with different interest groups. Once you have decided on the group and have brought them together, then the same requirement to keep them focused on the task applies. One method you can use is to get individuals to write on a piece of paper something that they consider to be the major problem or issue associated with the task. Collect the responses and list the items on a board or flip chart. It sometimes works well to get people to list single items on Post-its. Let people see what others have written and then make a second round, collecting any new ideas which may have been sparked off by viewing the comments of others. After several passes have been made and no new ideas are emerging, list the items and get the individuals in the group to rank order the items on the list. Display responses on a chart and use it as a discussion point by getting delegates to discuss the relative rankings. This technique can be used at the start of an analysis task in order to identify quickly problem areas or it can be used during the process to keep track of progress. It can also be used to generate items that can be used in a card sort exercise.

We can bring a level of quantitative analysis to the process by, once we have generated a list of items, asking group members to identify, in their opinion, the five most important points. Each individual is asked to rank their chosen items in order of significance. The most significant element is allocated five points, the next four and so on down to one point for the least significant item on the list. We suggest five items but you can allow more if your initial list is very long. However, the point being the maximum points awarded equates to the number of items to be chosen. Thus, if we ask for seven items, then the most significant is awarded seven points working down to one point as described earlier. Once the group has worked through their list, collect in the scores. Next, we can add up all of the points allocated to each item on the list and determine an order of importance. Because we can also work out a percentage distribution of scores, this technique allows us to visualise the contribution an item makes to the overall problem which, in turn, can be of use in deciding how to allocate resources.

Extracting Expertise

Control-of-Action, as we have already said, is the least considered aspect of training design, probably because investigation techniques have been lacking. However, over the past few years a loose collection of tools, generally known as Cognitive Task Analysis (CTA), offers the opportunity to get at people's control strategies. In this section we will consider a shorthand version of CTA which will still deliver results but without the need for complicated statistical procedures. First, though, we want to look more closely at this distinction between 'action' and 'control-of-action'.

After several accidents in 1998 in which a lack of pilot technical knowledge was cited as a cause, the International Civil Aviation Organisation expressed concern about the technical training of the pilots involved. A more relevant question, though, is did they lack knowledge or was it just poorly organised? During a study of error in air taxi operations in the UK, we found that 14% of incidents or accidents could be categorised as Knowledge-based errors (using Reason's[14] typology of error), 30% were Rule-based errors whilst the remainder were Skill-based errors (slips and lapses). A similar distribution of error types was found in a study of UK-registered Boeing 737s and Airbus A-320s. More importantly, in the Air Taxi study, 36% of knowledge-based errors resulted in death or damage to the aircraft as opposed to 8% of skill-based errors. Knowledge and rule-based errors can be traced to problems with the application of mental models whereas the skill-based errors are largely down to lapses in the attention of the pilots. These figures indicate that ineffective control-of-action resulted in nearly half of all errors recorded by pilots. Given that much pilot behaviour is routine procedural activity and does not make much call on problem-solving skills, in relative terms, we can see is that these error types are over-represented. Lest we offend our pilot readers, what we are saying is that skill-based errors, although forming the largest group numerically, do not occur as often as they could do given the amount of skill-based behaviour demonstrated by pilots. We can also see that knowledge- and rule-based errors more often result in accidents or incidents.

Having seen some evidence from real-world operations, we will now look at how an emphasis of 'action' at the expense of 'control-of-action' can distort training design. We did a study of the way military fighter crews set about shooting down a possible enemy. The task, as described in the manual, required the navigator, first, to set up his radar and then, by interpreting his radar picture, to talk the pilot through a particular flight profile. Eventually, with the enemy now in front of him, the pilot could release the heat-seeking missile at the hot tailpipes of the oppositions' aircraft. However, 2 things had happened since the course was first set up. First, the radar fitted to the aircraft was much improved and, second, active homing missiles (i.e. those that used a small on-board radar to track the target) were now fitted as well as the old heat-seeking versions. As part of the investigation, pilots were asked what they were trying to do at each stage in the attack profile. A very different task segmentation soon emerged. On receiving an indication on the radar screen of an approaching enemy, the first thing the crew did was to confirm that the radar return was, in fact real, and not some spurious blip thrown up by the technology. The problem with setting off to attack a piece of empty airspace is that you could reveal your hand to a real enemy who, up to that point, may not have known of your existence. Crews had developed a set of techniques - which included manipulating the radar - to accomplish that goal. The next stage was to determine what type of aircraft was represented by the blip. If it was a bomber, then your job was much easier. If, however, it was fighter then you had a different problem to solve. Again, crews had developed a set of manoeuvres and techniques that could gather the information needed to decide on the next stage, which was what type of attack to launch. Whereas the formal description of the task comprised a sequence of steps to be followed (such as; A. turn on radar, B. allow radar to warm up, C. set range etc), the real world task was more rooted in operational requirements and resulted in tasks being clustered to meet those requirements.

The applied interview technique that we have labeled as Cognitive Task Analysis tries to identify how skilled operators control activity. It also tries to distinguish between expert and novice performance. First, you need to identify a target population to be interviewed. They should all be previously trained in the task but you should aim to include some beginners as well as some experienced operators in the group. As in the standard approach to task analysis, get the individuals to describe the task in their own words. However, the difference here is that, instead of getting them to talk about what they physically do at each stage, we also want to get them to talk about what they are trying to accomplish at each stage of the process. We are trying to identify the goals to be achieved if the task is to be a success – in effect, we are trying to reveal the goal states discussed in the previous chapter. It is helpful to use lots of paper and draw things out as flow charts, again remembering that we are after processes as much as actions.

We are trying to break the task down into functional chunks. Once we have an acceptable, functional description of the task, go back to the beginning and get the candidate to describe, in explicit terms, what conditions they are trying to satisfy before they can move onto the next stage of the process. Get them to identify how they could tell when one section was complete and it was time to move to the next chunk of the task. We can double-check this stage by working backwards through the process and asking, at each stage, what needs to have been done before this part of the process could succeed. This may sound like over-kill but it is surprising how much information is revealed by getting people to look at their jobs from different perspectives.

The final stage involves getting delegates to identify how they could tell if things were going wrong. What were the signs they used to spot that things were going off-track. At this stage, we also need to get people to identify what methods, tactics, rules-of-thumb, etc they used to keep things on track. By then comparing the beginners with the more experienced operators, gaps in expertise are made clear and goals for training are starkly revealed. Experience shows that between five and eight interviews need to be conducted; little new information is gleaned from additional interviewees beyond this number. Again, the process is time-consuming with an average task taking up to two hours to explore fully. However, it should be clear that this type of analysis is best applied to those tasks that involve lots of 'head-work'.

From Task to Training Needs

By now, we should have produced a description of a set of activities required of our trainee in order to be gainfully employed. This is what we would, more properly, refer to as a Task Analysis. We should also have some idea of the set of contingencies most likely to be encountered once our trainee starts work. Finally, we should have started to tease out the various 'rules-of-thumb' or control strategies being used in order to move between the variations of the basic task which will invariably be encountered in the real world.

Another way of looking at this is to say that we have identified the performance repertoire, the skills needed to select the correct approach to a situation from the repertoire and, finally, the skills needed to adapt an approach to a novel situation.

Earlier, we offered as a criticism of training design that the process tended to stop at this point with little consideration of what trainees needed to learn in order to fully competent in their use of the performance repertoire. So, what else is needed?

As part of our analysis we will have identified some underpinning knowledge which supports skilled performance. In order to develop our Task Analysis into a Training Needs Analysis we need to make sure that we have fully explored this area of underpinning knowledge.

In many existing training systems we tend to avoid the issue in that we offer blocks of instruction which seem to be relevant to the area of activity. We then hope that linkages between the knowledge and skilled performance will either exist in the first place or will be made as training progresses. For example, most student pilots have extensive ground training in aerodynamics on the understanding that the knowledge acquired will assist them in understanding the performance of the aircraft in flight. However, we need to question the relationship between teaching aerodynamics and management of the flight process given the example of the 747 taking off from San Francisco cited in the Introduction.

Our complete Training Needs Analysis, therefore, requires a description of the underpinning knowledge that supports skilled performance. To a large degree, as mentioned earlier, this should emerge as we investigate the task. However, as part of our planning, we need to make provision for exploring this area.

The questions in Table 4.1 relating to objects used in the task, information required from others and the need to interpret documents, charts etc will give pointers to the need for underpinning knowledge. Similarly, the responses to questions concerning the control of activity should also reveal the requirement for such knowledge. The key concepts of a knowledge domain could be listed on cards and a Card Sort exercise used to establish the significance of subject matter.

Planning your Analysis

When planning your analysis, you need to bear in mind that no single technique will deliver all of the information you need to do the job properly. You should plan to use a mix of approaches, the mix being determined by the level of complexity of the task and resources available.

It is easy to under-estimate the time needed to do a thorough analysis. When planning an observational approach, remember that you will need to allow time to become known to the people who you want to observe.

Make full use of flow diagrams and other methods of graphically illustrating the way the job works. Look for meaningful ways of segmenting tasks and do not assume that the way the job is described in a manual will bear any relationship to how the job is actually done. In fact, one of the biggest problems is likely to be that of identifying units of performance. The CTA approach requires operators to chunk activity in some meaningful sense, based on the goal active at the time. Once you have finished the analysis, you will need some way of communicating the results, and that is what we turn to next.

Finally, you need to consider who will participate in the analysis. We have already described some of the difficulties involved in getting access to operators. In most cases, Subject Matter Experts (SMEs) will be appointed to answer all questions. When working with SMEs it is important to stress that the role of the SME is to proved expert knowledge about the job, not to tell the analyst how to design the training course. The SME undoubtedly has a broad experience of workplace failures and probably has a good idea of how they would like to see training conducted. However, at this point that is your job and not theirs. Take advice but beware of SMEs taking control.

How to Describe It - Constructing the Syllabus

Having conducted our analysis, we need a format for the output. The objective of the task analysis was to identify what people did and how they did it. We also identified the training needed in order to achieve the goal of producing a competent operator. Throughout, we should have deliberately tried to avoid configuring the analysis with some preferred training solution in mind. That said, it is not uncommon to find that the training analysis is produced as a justification for a prior decision to invest in an expensive piece of training technology. However, in the perfect world, we need to communicate the results of the analysis to those who will be involved in the next stage of the design process and we do this through a syllabus.

We also need to consider what other uses the syllabus serves. For example, in addition to influencing the design of training events, the syllabus will support bids for the purchase of equipment and it will influence the choice of staff. In its final version then, it can be a powerful document. For our purposes, we will concentrate on the specification of training requirements.

The most common way to define requirements, certainly over the past 30 years, has been through a training objective. Training analysis, as a field of activity, has its roots in the major US defence and space projects of the 1950s and early 1960s. The design, building and launching of nuclear-powered submarines, the training of personnel to operate them and the effort to put man into space and, then, on the Moon gave rise to a whole gamut of management techniques; training needs analysis was just one of them. The prevailing school of psychology at that time was behaviourist; researchers were concerned with inputs and outputs to the system and what went on inside the persons' head was out-of-bounds. Although behaviourist psychology has fallen out of favour, the training object remains as the primary way to define a training task.

Because behaviourists were concerned with observable behaviour, training objectives concentrate on things that people can be seen to have done. An objective usually has three parts; a performance statement, a statement of the conditions under which the performance should be observed, and a standard that must be met if the student is to be deemed successful. An example of a fairly simple example of a training objective is given in Table 4.2.

Quite often, a syllabus will contain only the Performance element of the objective. Table 4.3 contains two piloting tasks taken from the US FAA generic task analysis conducted under the AQP system.[15]

Table 4.2
A Simple Training Objective

Performance	Conditions	Standard
Recite the five times table from 1 to 10.	By rote, in a classroom, without access to a calculator or pencil and paper.	8 out of 10 correct.

The boxed letters in Table 4.3 stand for the type of learning required. In the behaviourist model of learning, the outcomes have traditionally been considered to be knowledge, skills or attitudes. Thus K is knowledge, M is motor skill and A is attitude. You may like to contrast these categories with the discussion of learning in Chapter 3.

The problem with the behavioural objective approach is that, for large and complicated tasks - like flying an aircraft - it can be tedious and time consuming to conduct a complete analysis. One task analysis for the First Officer position on a Boeing 737 consumed over 1000 pages of A4 paper.

Table 4.3
Example Objectives from the FAA AQP Generic Pilot Task Listing

1.4.2 Conduct Takeoff Briefing

1.4.2.1	[A]	Brief the type of takeoff
1.4.2.2	[A]	Brief the direction of turn after takeoff
1.4.2.3	[A]	Brief the initial heading
1.4.2.4	[A]	Brief the assigned altitude
1.4.2.5	[A]	Brief the destination and planned cruising altitude
1.4.2.6	[A]	Brief the abort takeoff procedures
1.4.2.7	[A]	Brief the emergency return procedures
1.4.2.8	[A]	Brief the ATC clearance

Table 4.3 Continued

2.5   Perform Takeoff from V1 to 1000 feet AGL

> 2.5.1 [M]   Remove hand from throttles
> 2.5.2 [A]   Call "VR" 3 to 5 knots prior to actual VR
> 2.5.3 [M]   Pull back on yoke to begin rotation of aircraft to charted takeoff pitch attitude
> 2.5.4 [K]   Observe pitch attitude on PFD and adjust pitch attitude accordingly
> 2.5.5 [C]   Comply with local or ATC climb restrictions
> 2.5.6 [C]   Observe appropriate bank limitations on departure
> 2.5.7 [A]   Call "Positive Rate of Climb, Gear Up" after observing positive climb established on altimeter, vertical speed and outside reference
> 2.5.8 [M]   Lift gear handle to the UP position
> 2.5.9 [K]   Observe EICAS gear up position
> 2.5.10 [M] Move gear handle to off position
> 2.5.11 [A] Call "1000 feet above"

An alternative to behavioural objectives is found in the more recent move towards defining competencies. A competence is a generic skill that can be transferred across areas of employment. A competence can also be defined as an element of work that a job-holder needs to accomplish if they are to be successful. Table 4.4 contains an example of a unit of competence taken from the UK Aviation Training Association's qualifications for pilots.[16]

Immediately, you can see similarities between competencies and objectives, at least in the way they are defined. The Performance Criteria is similar to the corresponding part of the behavioural objective. The Range Statement, at first glance, is similar to the Conditions element of the objective. However, in the case of the training objective, the Condition statement describes the conditions under which the trainee is expected to demonstrate achievement to the observer; in effect, the testing regime. With competencies, the Range statement is an attempt to indicate the contingencies a proficient operator would be expected to deal with in the workplace.

The major difference between the two is the statement of underpinning knowledge which tries to link theory and practice, an aspect which can be cumbersome to achieve in behavioural objectives.

Table 4.4
Example of a Pilot Unit of Competence

Unit A4:   Check and Operate radar and radio aids and carry out radiotelephonic communications
Unit A4.1: Check and Operate radar and radio aids

Performance Criteria	Range Statement	Underpinning Knowledge
A4.1.1 Radar and Radio Aids are tested for serviceability prior to flight according to laid down procedures	Radar and Radio Aids [list of types of equipment to be operated]	Basic Radio Principles; Theory of propagation of radio waves[etc]

The problem with behavioural objectives is that they are rooted in observable behaviours; you have to define what you expect to see before your eyes if a student was to demonstrate proficiency. Now, for much work-place activity, this is feasible. However, for much of the higher-level processing needed to make sound decisions, this would be impossible. In many cases, a correct solution can be the result of flawed decision-making: consider the example of using the FMS to manage the descent described earlier. The competence approach accepts that looser definitions are just as useful. However, the fact remains, we need a statement of performance.

The Conditions or Range Statement element establishes how we expect the person to perform. Whereas the behavioural objective states the conditions under which the testing of performance will take place, the competence range statement reflects the work-place situations in which the candidate is likely to perform. As such, this is a more useful piece of information. The FAA AQP analysis mentioned earlier requires air carriers to map training objectives onto the company's route structure and method of operation so that performance is tailored to work-place needs. Thus, although the examples of objectives given above were from a generic analysis, in reality even two companies flying the same aircraft should produce different sets of objectives to reflect their local operations.

As we said earlier, a useful addition found with competency movement is the attempt to determine any underpinning knowledge associated with the competency. The behaviourists would require this knowledge to be specified in behavioural terms, which is why this form of analysis can get quite messy.

Whatever method is used to define our training requirements, the fact remains we need something which can act as a mechanism for communicating to stakeholders what exactly we have set out to achieve through training. As a final word of warning, if the syllabus is defined at too high a level, its breadth will allow multiple interpretations. We will lose standardisation across instructors or training units. The result will be what is known as 'syllabus creep'. In the absence of clear guidance, trainers will tend to emphasise those parts of the course they either understand best or get most satisfaction out of delivering. For example, between January 1984 and July 1986 the British Army experienced 17 incidents in which damage to helicopters occurred during training in Engine-off Landings (EOLs).[17] During the same period, not one single event occurred in operational flying which required the use of this technique. On investigation, it was found that flying instructors liked practicing EOLs because they gave instructors the chance to demonstrate their superior flying skills to their trainees. On the other hand, if the syllabus is defined in too much detail, the training will be too specific and the development of transferable skills will be lost. Students will be unable to take general principles and apply them in different contexts; we will have failed to develop expertise.

Conclusion

In this chapter we have examined what is probably the single most difficult aspect of the training design process. Task analysis calls for attention to detail, access to the right people, can swing from high excitement in the analysis phase to mind-numbing boredom when you get to syllabus construction. However, an incredible number of training design projects start without a TNA having been conducted - and fail accordingly. The TNA serves many purposes and it is important not to confuse the issue. A description of the task, no matter how detailed, is of little use unless we understand how people do the job. We need to identify any problems encountered and strategies used to keep the system functioning smoothly. In the next chapter we will look at ways of measuring effectiveness.

Case Study

Imagine that you have been called into a meeting with the aircraft Fleet Managers and the Director of Cabin Services (or whatever titles are appropriate in your organisation). They tell you that there have been large numbers of complaints from pilots and cabin crew about the Rostering and Crewing sections. The roster is often published late and is sometimes inaccurate (crew who have reported sick are rostered to fly, agreed days off are ignored) and very demanding to fly (too many consecutive early starts). Crewing always seem to be calling people out at short notice, their telephone manner is considered rude and, sometimes, their solution to a crew problem seems to cause enormous disruption to the roster. People employed in these sections have, traditionally, learnt on the job. They come from a variety of backgrounds and, often, have no prior experience of aviation. You are asked to advise on the design of a training course for the Rostering and Crewing sections. Devise a TNA plan.

5 Exams and Testing

Introduction

In the training design cycle described in the Introduction we saw that the design of test items came after the initial training needs analysis and before we started work on course design proper. This is because any training process requires you to have a clear idea of the end-point. We need to know the operational performance to be achieved by students once they have finished training. We need to be certain that graduates from our course will be capable of useful work as a result of our efforts. Therefore, a testing regime must be established which will tell if a candidate has reached the end-point. After that, you design the events required to allow the students to achieve the desired level of performance by the end of training.

In this chapter we will look at testing in more detail. At the end of the chapter you will have:

- Considered why we test students.
- Identified some problems associated with testing.
- Looked at how to test different types of performance.
- Designed some test items.
- Considered what to do with the output from the testing process.

Why Test?

We said earlier that testing allows us to check if students have reached the graduation standard. But there is more to it than that. Information from testing can provide feedback to the organisation about the success of the training system. Examinations represent a snapshot in time and, so, allow us to gauge the progress of students. Answering questions helps students to consolidate their learning; it also directs them to the more important parts of the syllabus. The output from examinations can be used to diagnose areas of misunderstanding on the part of students and can, therefore, be used to direct remedial activity.

If we concentrate on the relationship between learners and tests, we can identify two types of information which can be derived from testing;

> Formative information and
> Summative information.

Formative exams are designed to help to remedy weaknesses in students. Entry-level testing tries to determine what students already know and, therefore, what they need to be taught during the course. Progress Tests are designed to establish how the student is doing so far and whether there is any need for remediation. The emphasis is on feedback to the student in order to facilitate further development.

Summative testing determines the extent of learning. It attempts to establish an overall measure of accomplishment. From the description of testing at the start of this chapter it could be argued that, at this stage, we need only be concerned with summative evaluation as we are only interested in whether the student has reached the graduation standard or not. However, like most things in training design, testing is maybe not as straightforward as it might seem. To ignore the double-edged nature of testing would be to miss an opportunity to build an effective training system, and so we will discuss both aspects in this chapter.

As an example of the interplay between formative and summative testing, look at this extract from a discussion between a tutor and a student. The subject is meteorology; in particular, how uplifted air can give rise to rainfall.

Tutor: What happens when moist air blows over Washington and Oregon by those winds?

Student: It condenses and it rains.
Tutor: Yes, why?
Student: Because the moist air cools and the clouds can't hold the water and so it rains.
Tutor: OK, what causes the moist air to cool?
Student: It cools when the wind blows and it lowers from the sky.
Tutor: What happens to the temperature of moist air as it rises?
Student: It gets warm.
Tutor: No, warm air rises, but as it rises it cools off. Do you think the mountains in Washington and Oregon have anything to do with cooling the moist air blowing off the coast?
Student: Yes.
Tutor: How?
Student: As it rises, or even if it doesn't rise the cold mountains could cool it off.

From this conversation[18] we can see that a test item designed to check if the student could state the weather associated with warm air masses blowing onto the coast of Washington and Oregon, summative testing, would elicit a correct response. We can also see that a further question aimed at determining the immediate cause of this weather effect would also get the correct response. However, as the dialogue develops, we can see that the student is not too certain about why these things happen. The effect on temperature of lifting air masses, the role of the mountains as agents of that lifting and the direct cause of cooling are all imperfectly understood, if at all. Thus, formative testing, at this point, would aim to determine flaws in students' understanding.

We have already referred to testing as a means of feedback to an organisation on the success of the training system and we will make reference to the point again in Chapter 7. In the meantime, however, here is an example of a problem which could be thrown up by using an examination system as a quality measure. A course for Air Traffic Controllers was repeatedly experiencing failure rates of 100%. At the time, the fault was considered to be the decline in the calibre of applicant and, hence, the current students lacked the qualities needed to cope with the course. The measure of performance used was final examination scores. However, on investigation it was found that several other factors were involved. The technology in service had progressed faster than the training system could keep up. The cadre of instructors had become progressively more junior and less experienced. Thus, their ability to develop the students on the course was minimal. The job, itself, was under threat and the school was being used as a means of recruiting exceptional candidates who could defend the interests of the profession as well as do the job of controlling aircraft. In this situation, a simple relationship had been established between exam pass rates and quality of recruit. In fact, no such correlation existed in reality. However, the symptom - poor exam performance - was indicative of more fundamental problems within the training system. We will now look at some problems associated with testing in more detail.

Problems with Testing

Product versus Process

We need to remind ourselves of the original purpose of designing training in the first place. We said that we wanted to develop expertise within our workforce so as to maintain our competitive edge and to allow people to contribute to our business goals. We need people who are, of course, competent at what they do but the best performers are more than that, they anticipate problems and solve them before the problem becomes either a cost or a hazard. We can identify two components of performance;

> Product - what is observable at the end of activity.
> Process - the method by which the product is achieved.

By product we mean, for example, the results of a calculation; the recall of a definition; the production of an accurate representation of an object of interest. Products can be completed weight and balance forms, a flight plan, the correct calculation of heading and speed for a navigation exercise, the calculation of the range of a radio aid using the relevant formula and so on. If we look at the area of soft skills, a product could be 'Recalling the definition of Latent Error', 'naming some representative dimensions of Personality' or 'Communicating Effectively'.

When we get to process, things get more difficult. For a start, a product is exactly that; it is something the student has generated which we can then observe. Processes, on the other hand, are the internal reasoning by which the product was arrived at. If we add two numbers together, the answer is the product but the working out in our head is the process. Processes occur internally and therefore are not so easily observed. Processes require some interpretation on the part of the observer. We need to differentiate, here, between the earlier discussion of action and control-of-action and the current topics of product and process. Although, superficially, they may seem the same, in fact the action/control-of-action pair gives rise to both process and product.

Occasionally, products will be generated at interim stages in the process. Thus as the student calculates the performance of the aircraft she will need to refer to particular documents. Selecting the correct document is, in itself, a product. She will need to extract data and manipulate them in some way. The outcome of that transformation will be a product. However, her understanding of the performance of the aircraft in the air, the relationship between formulae and behaviour of the aircraft and her preparedness to control that aircraft in flight are all things which, to a large degree, are hidden at this stage.

In reality, we all know that we can get the correct product from an incorrect process. In the real world, process is what we pay people for because, quite often, there will be many acceptable solutions to a particular work-place problem. In fact, from our earlier examination of expertise, selecting the appropriate solution that will fix a problem at that particular moment, given prevailing circumstances, is probably the most important skill we need to develop. In training, we find that it is easier to measure product than process and so we need to be concerned with how our testing strategy will deal with the importance of process over the ease with which we can measure product. This relationship between product and process is intimately tied up in our selection of testing methods.

If we return to our examination of learning outcomes in Chapter 3, we can see that testing for propositional knowledge will be much simpler than testing for effective mental models. To an extent, this was illustrated by the student and tutor discussing rainfall on the United States Pacific coast. In the context of the product verses process debate, the method by which the student arrived at the answer to the question 'what happens when moist air blows over Washington' could be considered irrelevant. This can be classified as a piece of propositional knowledge that is stored in memory and retrieved in response to the appropriate question; the Product is all. However, in trying to promote understanding of the subject, we need to be sure that the student understands what causes the rain to fall. Again, we can build a chain of propositions thus: Warm, moist air hits coast - air rises - air cools as it rises - as air cools, rain falls.

Each of these propositions can be stored in memory and recalled as appropriate. By testing for product we can draw some conclusion about the student's level of understanding and, in this case, we could probably be convinced that he does, indeed, know what is going on. The student's learning would, though, be more robust if they had an effective mental model of the atmospheric processes involved in this case. Therefore, testing for process, which is what the tutor is doing throughout the extract, is probably more important. What we can also agree on is that the methods by which you test for each are likely to be different. In simple terms, it is probably fair to say that questions starting with 'What' will elicit propositional knowledge while those that start with 'How' or 'Why' will get closer to testing for comprehension.

Validity and Reliability

The next problems we need to overcome concern the extent to which our testing methods are Valid and Reliable. A valid examination actually tests what we think we are testing. Imagine that we have asked a student pilot to recite the formula for the lift force acting on a wing. If his response is correct, have we actually tested his ability to remember by rote or can we assume that he understands how lift is generated? What we are trying to achieve is a training system where good performance in an exam will correlate with good performance in a related real-world task.

Reliability is the extent to which a student's exam results are consistent over time. If reliability is high, then exam candidates will get similar marks if re-tested after a period of time. What we want to achieve is a system where we can have confidence in the scores obtained. Thus, a high score in the exam is a reflection of ability and will be indicative of a better performer than would a low score. After all, the aim of the training system is not to produce people who can do well in training; we want them to do well in the real world. We need confidence that someone who does, indeed, do well on the course will also do well once they start work. Many courses are of short duration and may only contain a single examination. Thus, we need to be confident that the score attained in a single exam is a reflection of ability and not chance. We will look at the use we can make of exam scores later in this chapter and we will return to the relationship between performance in training and in the real world when we discuss the evaluation of training in Chapter 7.

Standardisation

At an operational level, running testing systems will involve ensuring that all candidates are tested in a standard manner. This is less of a problem when we use standardised methods such as multiple choice objective testing. However, when we use more flexible performance-based methods then standardisation can become an issue.

We also need to be certain that our testers are equally standardised. Inter-rater reliability, as it is known, can be a serious issue, especially in subjective areas of training. Here is an example of what we mean from basic pilot training. A video camera was mounted in the cockpit of a single-engined light aircraft. The field of view was sufficient to see enough of the outside visual reference as well as the instruments and most of the hand movements of the trainee pilot. The student was recorded making a series of take-off manoeuvres. A group of experienced flying instructors was assembled in a room and shown the videos. Each instructor had to assess the student's performance using the normal reporting formats, which included a numerical score and narrative comments. After the exercise, the instructors' reports were collected and all of the numerical scores put on a single matrix. Needless to say, there was considerable variation between instructors. The narrative comments threw some light on the reason for the spread of scores. One instructor wrote 'a poor take-off' while another wrote 'not bad for this stage of training'. Although both instructors viewed the same performance, they each used a different frame of reference. It was, indeed, a poor take-off but equally, given the student's inexperience, it was not bad for the stage of training. Given that much of what we want out of our training systems will be real-world behaviour as opposed to success in a pencil and paper test, the matter of inter-rater reliability is important. The answer is to have clear guidance concerning levels of acceptable performance, clear indications of what to observe and, finally, frequent standardisation meetings, including exercises like the one just described.

Shallow versus Deep Processing

One of the problems with training designed to achieve success in an examination is that it results in the accumulation of 'superficial, rote knowledge'.[19] That this claim can be levelled at the aviation industry, at least in the UK, is illustrated by this story from one of the UK CAA examination team who changed the wording of a question in the professional pilot exam.   Despite the fact that the test item remaining constant, the failure-rate on that question soared.   The problem is mirrored in the USA where students can pass the written examinations but still fail the, more flexible, oral conducted by the flight test examiner.   The problem is caused by students trying to learn the surface features of the question rather getting to grips with the underlying knowledge.   This is a reflection of the Shallow Processing discussed in Chapter 3.   Our aim should be to promote deep processing in which effective memory structures are built which lead to genuine understanding.   Our testing regime will play a part in that process.

How to Test

We have, throughout this book, been coming back to the twin ideas of competence and expertise.   At the end of the day, we need someone to perform in a manner which is effective and safe.   We can try to break down this competence into the microscopic learning outcomes but, in reality, it is the combination of these elements which is what we call competence.   In school we have become accustomed to testing as a periodic recitation of stored facts and the model is still prevalent in aviation.   Thus, if we sit a ground course there needs to be a test.   At some UK commercial flying schools, more students fail the ground exams than fail the flight tests and, yet, the ground theory is supposed to support the airborne exercise of piloting skills.

The measurement of this accretion of learning to form skilled performance is the major problem we have to overcome when designing our testing regime.   The most valid and reliable test would be to ask the student to accomplish a real-world task which contained all of the key performance elements.   However, these are difficult to construct, expensive and time-consuming to run.   Therefore testing has to be a compromise between meeting the need to confirm that a graduation standard has been achieved and the resources - time and money - available to conduct the tests.   We also said earlier that, of the 2 components of performance product was easier to observe than process.   So, with all this in mind, what types of test are available to us?

Open-ended Tests

Perhaps the most familiar test format for most of us is the essay. Favoured in school systems for many years, writing an essay has become the traditional form of end-of-course assessment. The advantage of an essay is that it enables us to track a student's train of thought and, so, go some way towards revealing process as well as product. The disadvantages of using this exam format are that they are extremely time-consuming to mark, suffer from problems of assessor standardisation and, perhaps most important, can disadvantage students with poor written communications skills. Open-ended tests also allow the student to hide a lack of knowledge by side-stepping the question and answering something slightly different. In fact, generations of students have been exhorted to Read The Question in an attempt to ensure that the answers given match the intended question.

An alternative to writing full essays are those exam formats which require students to provide short-answers or to fill the gaps in prepared statements. This approach reduces the risk of students' success being dependent upon the ability to write. It also reduces the burden of marking exams. However, there is still the risk that a student could fail the exam because of a misunderstanding of the question as opposed to a real lack of knowledge.

Practical Tests

Open-ended testing methods have a part to play in assessing competence and part of our training design strategy involves choosing the most appropriate testing regime. However, we are also interested in evaluating work-place competence. Therefore, it seems sensible to test skills in a more realistic setting that can be achieved with pencil-and-paper tests. As part of our testing, we could ask students to demonstrate sub-sets of the complete performance. For example, we can get students to use relevant documents to prepare a flight plan or an aircraft trim sheet. At the next stage, we can get students to work through more complex parts of the task in order to demonstrate competence. Thus, we can get student pilots to fly a particular approach at an airport using a flight simulator. We can get a cabin attendant to demonstrate the routine for preparing the cabin for boarding using a diagram of an aircraft. We can also test skills using a false task. For example, an experiment that looked at training transfer found that success in a computer shoot-em-up game improved performance in instrument flying for pilots.[20]   Thus, we could use the same task as a reliable test of instrument flying skills.

The discussion, so far, has been around testing in basic training. However, if we consider the wide-spread use of recurrent training, we can see that we also need testing systems which fit into this form of training. One approach adopted in the domain of crew resource management is that of having behavioural indicators of acceptable performance. Table 5.1 lists some examples of behaviours an evaluator, or 'tester', would look for when observing a crew at work.

Table 5.1
Examples of Behavioural Indicators

'Preparation and Planning
    Crew plans ahead, monitors developments, anticipates required actions.
    Key decision points and bottom lines established
    Shared analysis of what's happening.'

For pilots, we can start to consider the automation of this process. With the introduction of digital Flight Data Recorders and the availability of Quick Access Recorders and suitable software, it is now possible to monitor pilot performance in a routine manner providing a form of continuous assessment of line flying. Hand-held computers loaded with databases of performance indicators now make work-place assessment of almost all groups of staff simpler and more convenient than ever before.

A problem with practical forms of assessment is that it is difficult and costly to set up. Furthermore, it is difficult to establish a standard test across all students and, again, inter-rater reliability is a problem. However, there is an intuitive sense of practical assessment as being a more valid and reliable predictor of ability than pencil-and-paper testing. This may be true, but it does not deny other forms of testing a place in our scheme. We will move on to this aspect later but, first, we need to consider the final category of tests.

Closed Tests

Unlike open-ended tests, where the student has a broad scope within which to answer the question, closed tests present the student with a range of options from which he selects the correct option. Students can be asked to decide if a statement is true or false, they can label objects on a diagram or match objects. One of the most wide-spread forms of closed testing found in many areas of training today is the Multiple Choice Objective Question (MCOQ).

The almost ubiquitous MCOQ was designed as a way of overcoming most of the weaknesses previously described for other testing methods. The format is standardised across all students being examined, does not rely on levels of literacy as a pre-requisite for entry to the exam, can be easily marked and has no inter-rater reliability problems. Despite these supposed advantages, MCOQs are much-abused in training. They are difficult to write, are vulnerable to students guessing the correct answer and can give rise to the shallow processing. There is also a risk that they will cause students to learn the wrong answer. For example, a student selecting the wrong option and then not receiving feedback on performance runs the risk of never knowing what the correct answer to that particular question was. Because of their wide-spread use, the design of MCOQs will be considered in more detail in the next section. What we now need to consider is how to select the appropriate testing system for our course.

Performances to be Measured

The first thing to remember when setting up the testing regime is that our goal is to determine if the student has reached our desired end-point; in effect, are they now safe to enter productive employment. Later on, in Chapter 6, we will look at how economics may require us to send out part-trained students who will finish their development whilst in-service. For now, we will take our end point as being a competent performer. This assumes that, from our Training Needs Analysis, we know what the required level of competence is. Furthermore, we should have defined the required performance as a set of objectives. We now need to review the types of learning we outlined in Chapter 3 and see what tests we can set up to measure performance. Some suggestions are contained in Table 5.2.

Designing MCOQs

Design of Questions

Because the MCOQ[21] is widely used in aviation, we will look at this type of test item in more detail. There are three component parts of an MCOQ;

The Instruction
The Stem - posing the question
The Responses - possible correct answers

The Instruction describes the mechanics of the question and should contain details of how to respond. It should also contain information about how scores are allocated. For example;

> Each question is followed by 4 answers of which only one is correct. You are to select the answer you believe to be correct. Marks will be awarded for a correct response and will be subtracted for an incorrect response. You will gain 2 marks for a correct response. You will lose 1 mark for an incorrect response. Unanswered questions will be awarded zero marks.

A single Instruction can be defined at the start of the exam or you may need to provide variations according to how you have designed the test item.

The Stem is the question proper. It defines the problem to be set the student and poses the question to be answered. For example;

> For a warm, moist air mass blowing onto the Pacific North West Coast, which of the following factors will cause rain to fall.

The Responses are the proposed answers from which we want the student to make a selection. Traditionally, MCOQs have offered 4 responses. There is, in fact, no rule that states an optimum number but the implications of the number of responses offered will be discussed later.

Table 5.2
Test Items for Different Learning Outcomes

Learning Outcome	Type of Test
Propositions	True/False Object Matching Free Recall from Memory of Components (facts, concepts, etc)
Schema	Free Recall from Memory of Structure (This is an example of x because...) Sort Items into Categories. Classify Objects Sentence Completion
Mental Models	Predict Outcome
Rules	Trace Activity (What happens if...?) Select Rule (How do I...?) Apply Rules to New Domain (If this is true here, then this should happen in this new case)
Skills	Perform or Cause to Happen Construct Proof Complete Task A while attending to Task B

We said earlier that MCOQs are notoriously difficult to write. They require the designers of test items to have a thorough understanding of their subject. Training establishments that use MCOQs frequently have to rewrite exam questions because a student has found another flaw in the framing of the responses. In a survey we did of the draft questions for pilot testing in Human Performance and Limitations for the Joint European License, of 292 questions submitted, 131 (44.86%) needed redesigning. The types of flaw observed in the questions are summarised in Table 5.3.

Nine of the questions in the proposed bank featured in more than one category of error. The questions deemed lacking in discriminatory power were considered too trivial or insignificant for inclusion in a professional pilot license exam. The sundry group included questions using jargon or units of measurement which were not in common use in all countries subscribing to the scheme. Some of the questions were more statements of belief and, in other cases, the accuracy of the answer was in doubt.

Table 5.3
Major Categories of Error in MCOQ Design

Category of Error	Number of Questions (%)
Wording of Stem	44 (15 %)
Wording of Responses	32 (11 %)
Relevance to Training Objective	23 (8 %)
Discrimination between Students	20 (7 %)
Sundry	17 (6 %)

Marking MCOQs

This data demonstrates some of the pitfalls in MCOQ design. One of the other problems mentioned earlier is that it is possible for students to guess the correct answer. We also mentioned that it has become traditional for MCOQs to contain four responses. Therefore, there is a one in four chance of students simply guessing the correct answer. In the illustration of the wording of the question stem given earlier, reference was made to the possibility of losing marks. The concept of penalty marking, i.e. subtracting a mark for an incorrect answer, was developed as a means of combating guessing. If the student did not know the answer, then the less risky strategy was to leave the question blank. To attempt to guess could result in the loss of marks if the guess was wrong. As an example of stakeholder power, penalty marking was dropped from UK pilot exams in the mid-1990s after the major professional pilot schools complained that too many of their overseas students were failing the exams. The high failure rate was damaging the prospects of winning contracts from foreign airlines. The actual purpose of the penalty marking system seems to have been forgotten in all of this.

An alternative to penalty marking is to factor a student's exam score to allow for guessing. This can be achieved using the following formula:

$$\text{True Score} = \text{Number Correct} - \frac{\text{Number Wrong}}{(\text{Number of Responses} - 1)}$$

For example, supposing the student achieved a final mark of 85 correct and 15 incorrect answers, with each question having four responses. Their true score would be:

$$85 - \frac{15}{3} = 80$$

Enhancing MCOQs

Before we move away from question design, we need to remind ourselves that MCOQs are not end in themselves, the simply represent a structure within which we can examine students' learning. If we look at Table 5.2, we can see that the test items suggested for propositions, schema and rules all lend themselves to the MCOQ structure. On the other hand, skills and mental models may be difficult to capture using this method. By adding some additional responses to the MCOQ format we can start to move towards the testing of more fundamental knowledge. We need to consider the use of such responses as:

a. None of the above
b. All of the above
c. Insufficient Information

Response a. is intended to force the student to actually read the question and understand it before answering. It aims to deter students from reading the responses first and then trying to match the most appropriate one to the question, a common student tactic in MCOQ exams. In effect, the student must first create an answer and then check to see if it is included in any of those listed.

Response b. can identify the faulty use of rule-based behaviour in that, in some situations, there may in fact be several correct solutions. Response c. forces the student to effectively analyse the problem before answering and can be used to identify schema, rules and mental models.

Although we have now extended the use of pencil-and-paper testing into broader realms than the simple recall of propositions, we also need to be aware of the fact that, to fully test performance, we need to provide a mix of testing situations, including practical observation of performance. We discussed the use of behavioural performance indicators earlier and the Repertory Grid technique described in Chapter 4 is a useful tool for deriving performance standards. These, again, are forms of testing which suffer from all of the drawbacks described above. However, although the effort needed for a sound testing regime should not be under-estimated, the fact remains that we need effective tests if we are to be able to have confidence in the output from our training courses.

Planning a Testing Regime

We can now see that testing is, perhaps, more complicated than originally thought. We need a clear plan to make sure that the performance of trainees at the end of the course is what we intended. The first goal of the testing regime is to make sure that we evaluate performance against the full set of objectives. As an example of the problem, we studied a course designed to teach airport staff about handling dangerous substances. The course had 37 declared objectives of which 14 were not tested during the end-of-course exam. The exam comprised 15 MCOQs. Two of the questions covered material from three different objectives and a further four questions covered two objectives each.

We have already made the point that testing needs to be valid and reliable and we have also made the point that not all aspects of performance are amenable to formal testing. Therefore, our testing regime needs to be carefully thought through. In the next chapter we will consider how to build courses and one aspect we will cover is how the curriculum can be driven by real-world tasks. The implication of designing courses around actual tasks is that higher-level performance i.e. the completion of a job element, should depend upon mastery of lower-level job elements. To pick up on an example used earlier, if the student can correctly compile a flight plan then we can have some confidence in the students' understanding of airspace regulations, use of charts and documents, and so on (although remember the product v process indemnity clause!). We can extend this principle to the design of tests. From our analysis we can identify aspects of a topic area where mastery at a particular level depends upon a thorough understanding of the underpinning knowledge. For example, returning to the Pacific North West, we could ask our student what would happen if, in another circumstance, the air mass hitting the mountains was particularly cold or dry. The correct answer would depend upon the effectiveness of the student's mental model as opposed to recall of propositions. We can develop this concept to construct a hierarchy of testing, illustrated in Figure 5.1.

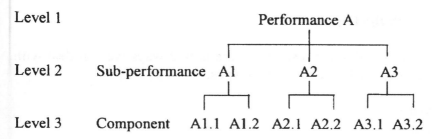

Figure 5.1
Planning a Testing Strategy

As a result of our analysis we may decide that Components A1.1, A1.2 and A3.1 will be tested individually. We may also test Sub-performance A2. In so doing, we can accept that Components A2.1 and A2.2 have been accomplished as success at A2 level would not be possible without mastery at the component level. Finally, by testing at Level 1, the Performance level, we can be satisfied that all other elements have been accomplished. The distinctions made here between performance, sub-performance and component and Levels 1, 2 and 3 are purely for illustration. Our intention is simply to reinforce the fact that testing requires as much planning and thought as the rest of the training design process.

Much testing is conducted at fixed points during training or at the end of the course. In many cases, progress tests are set not so much to provide formative feedback but, instead, to act as hurdles to decide on access to the next stage of training. We can afford to be more flexible in the use of testing methods, finding ways to introduce progressive or continuous assessment that contribute to learning.

In a commentary on education in the USA a UK news magazine,[22] talking about testing, observed that 'students simply checked off answers on multi-choice lists; a test of memory or luck but hardly of coherent thought'. In work we did for the UK CAA, we found that the majority of questions in the Commercial and Air Transport Pilot ground exams called for simple recall of propositional knowledge as opposed to applying knowledge to solve problems. The testing regime you design needs to support student learning whilst confirming achievement of objectives. Once we have designed our tests, there is one more aspect we need to consider; what do we do with the results?

What to do with the Output

Having put students through tests, there are now 2 ways we can deal with the results. Exam results can be of two types:

Normative
Absolute.

Normative results compare students with their peers in order to produce a course rank order. In effect, the exam results are used as part of a selection process in that students will be disposed of at the end of the course according to their position on the list.

An absolute form of reporting simply compares performance against a benchmark. Thus, students either passed or failed. Grades of pass may be determined if there is a purpose to be served. But this is not all we can do with exam results. We have already discussed the need for test items to accurately assess learning and, in our discussion of exam design, one of the observations we made was that some questions seem to be poor discriminators between students. As part of an investigation into a course to convert pilots to a new aircraft type, which will be explained in more detail in Chapter 7, we examined the questions used during the ground school phase.

The phase tests and final examination attracted some adverse comments on the course evaluation forms; 'more an interpretation exercise than a test of knowledge', 'sometimes as much a test of English', 'wording very ambiguous'. These observations underscore our earlier comments about the design of questions. The testing system comprised five phase tests, each of 20 questions, and a final test of 114 questions. Given that the course was 12 days long and included many hours of computer-based training as well as familiarisation exercises in a simulator, this is a fairly sparse sampling regime. Students' understanding of the aircraft's electrical system, for example, was covered by five questions in one of the phase tests and 6 in the final test. Flight Guidance was tested by 12 questions in the phase tests and 6 in the final exam.

The exams were exclusively MCOQ with four possible answers. Suppose, for a moment, that four candidates all randomly selected an answer from the four on offer. As we have already said, it is possible that one of the students could get the correct response purely by chance. Now, the implication is that, for any question, if 25% of candidates get that question correct, we cannot say anything other than the effect could be pure chance. It says nothing about the validity of the questions as tests of knowledge. However, if less than 25% get the question correct then we need to ask if the training has been inadequate, if the question is a good discriminator between candidates or if something is wrong with the design of the question. For the course we looked at, seven questions from three of the phase tests and 12 questions from the final exam fell into this category.

Now let us consider the opposite case. If a test item is valid then we can rely upon the fact that a candidate who gets the question correct does, in fact, understand the subject. However, in practice, a question which achieves a 100% pass rate is either too easy or it tests something which is too insignificant to be a valid or reliable indicator of mastery. Of the test items examined, 15% fell into this. Although no absolute cut-off figure exists, we can argue that any test item that achieves a 95% pass rate should be considered of little value. By this criterion, the number of dubious questions rose to 38%, or 42% if the questions discarded for excessive failure rates is included. In short, given that these exams could, and were on one occasion during the study, be used to suspend a pilot from training, on closer inspection we would have to have doubts about nearly half of the items in the question bank. So, even after we have designed the exam, we need to rigorously evaluate its fitness for its intended purpose.

Conclusion

This chapter has been littered with warnings about the pitfalls that can befall constructors of tests. Therefore, we will end with another. As this chapter was being written the author caught a speaker on a radio programme discussing the British education system who said 'one of the problems is that we start off by making the important measurable and end up making the measurable important'. No better words of wisdom could be offered at this stage. We have seen that no one form of test is foolproof. They all have drawbacks, be they administrative or in the sense of effectiveness as measures of performance. If we go back to the discussion of product and process, there are some aspects of a student's performance which are easy to capture while there are others which are almost impossible to observe. It is our task to make sure that we test what is significant from the point of view of the job.

There are many areas where flaws in our training design can have an effect on our examination system. To close this chapter, we summarise some of them here:

- Defining the course content - if the content is wrong then our testing is invalid.
- Selecting the item to test - if the test item is not related to the task then, again, the testing is invalid.
- Writing the question - poor performance in the test could be because the question is poorly worded, not because of a lack of knowledge on the part of the candidate.
- Student interpretation of the question - does some aspect of the student's background cause an idiosyncratic interpretation?
- Students' responses - does the question call for response type that is beyond the candidates' ability even if candidate knows the correct answer?
- Response evaluation - is the system of marking unbiased, agreed, standardised and objective?

Case Study

Crew Resource Management has probably been the major growth area in crew training in the past 10 years. Increasingly, the lessons of CRM are being applied to other work groups such as maintenance, dispatch, ramp workers and so on. The subject matter deals with some of the softer, social skills but, as we have seen, there are moves to include the evaluation of these skills on the line, at least for pilots. Imagine that you have designed a CRM course for your company. One of the subjects you covered was 'effective team-working'. You now need to consider how you will test each individual on the course to confirm that your course has contributed to their team-working skills. How are you going to construct your tests? You may want to consider these aspects:

- The identification of behavioural standards.
- The methods of testing.
- How to establish the validity and reliability of your tests.

6 Curriculum Building

Introduction

This chapter will deal with the problem of organising learning events into a coherent course. We need to consider how we are going to match events to the learning goals we are trying to achieve. And we must give some thought to the medium of instruction we plan to use. It is amazing how often an automatic assumption is made that instructor-led, classroom-based activity will be the preferred method even before any analysis of requirement has taken place. At one airline, for example, we asked what alternative methods of instruction had been considered. Specifically, we were looking at the suitability of computer-based training for pilots, given that the airline only recruited in small numbers and infrequently. The training manager responded with 'the Chief Pilot likes 'chalk and talk' so that how we train'. Building the course can be the most interesting phase of the training design cycle - and it can also be the most tedious. At the end of this chapter you will have:

- Considered the need for a clear course aim.
- Identified different methods of sequencing learning events.
- Examined the relationship between learning and the events of learning.
- Considered different media of instruction and methods of delivery.
- Consider the factors affecting the selection of media and methods of delivery.

Establishing the Aim

It probably seems a little back to front to discuss the course aim at this point. In the chapter on task analysis we looked at methods of specifying what people needed to do to be successful in the workplace and in the last chapter we considered ways of telling if students had achieved that goal. So what else is there to say? To answer that question, let's return to the 2 experts who we saw working on the DFGS task in Chapter 3. In an attempt to identify what, exactly, the course was about, we spent half a day trying to establish the overall course objective. Here is what we came up with:

> 'The aim of the course is to provide ground instruction which will enable a student to commence simulator training with a sound knowledge of aircraft systems. A sound knowledge implies:
>
>> An ability to describe the significant behaviour of an aircraft system.
>> An ability to describe any relationships between systems.
>> An ability to describe the implications for the safe operation of the aircraft of a loss of a system or sub-system'.

Immediately, we can see that the phase of training in question is in preparation for another phase: the simulator. So, we have an upper limit, so to speak, of what we are trying to achieve. Of course, we would also like to know what extra training was going to be delivered in the simulator so that we could be certain that nothing was missed. It also provides us with some criteria we can use in deciding if a piece of subject matter is relevant. In terms of the level of expertise we are trying to develop in students, as training designers we could probably do with some amplification. From the student's point of view, it is probably an acceptable statement of why we are all sitting in this classroom... or reading this book... or sat at this computer terminal. From your perspective as the designers of training, you need to identify the goal you are trying to achieve so that you know when to stop, which is why you need to spend some time trying to develop an overall course aim.

In fact, what we have achieved with this exercise is to make the first step in distributing training requirements across a range of activities; in this case, ground-school and a simulator. Our TNA will have provided a broad description of needs but which must now be focussed in some way. We need some way of applying resources to a unit of performance which is attainable within the resources available and is appropriate to the needs of the job.

In the case of the aim described above, we sat down with the head of the training organisation who was also an expert in the sense that he was a skilled instructor on the aircraft in question. After some initial discussion of the overall objectives of the course and problems with the existing method of training, we began to focus on what was required of the planned new system. The expert was asked to state the performance expected of successful graduates from the course. For each statement, the expert was asked to give an example of what they meant by, for example, 'a sound knowledge'. In effect, how could someone 'see' sound knowledge. Each statement was tested by justification; the expert had to say what the outcome would be if the student did not perform in the way specified. If a consequence could not be identified then the statement was discarded. Through a process of iteration and distillation, testing and refining, the overall course aim was arrived at. And then the work began.

In some ways, establishing the overall aim is an insurance policy. The methods used in its development and the form of description are more or less the same as those used in the earlier analysis phase. However, task and training needs analysis alone will not establish a course aim; it is not the purpose at that stage. Without an aim, courses can wander off track or get bogged down in trivia. With our aim clearly in mind, we need to start mapping out our approach to the course. We have two specific questions to answer: what are we going to do and what order are we going to do it in? The first question can be further expanded to consider the course content and the course activities that will enable students to come to terms with the content. We will deal with the selection of content in the next section.

First, though, we need to address two other problems. The first is who will attend our course and the second is at what point do we stop training. The first question deals with the input standard of the students. We need to have an understanding of what our course members already know. If we return to the course aim described above, we can assume that prospective students can already fly, will have finished all the exams associated with gaining a pilots licence, may well have experience on other aircraft types. All this will have a bearing on the starting point of our course. The second question is really one of resource allocation. Of course, we would like to train to proficiency; our students will be fully capable on graduation of doing the job. However, this goal may not be attainable. We may not have the time or opportunity to train to proficiency. However, we need to be aware of where the cut-off point will be. As part of our strategic planning we need to have an idea of how the gap between graduation standard and proficiency will be bridged. We need to consider these questions in more detail as we start to work on building the course.

Selection of Course Content

In Chapter 3 we talked about levels of description in relation to defining the syllabus. In Chapter 5 we talked about testing as important in deciding is students had achieved our course goals. However, as we have just said, still need to consider how much training to deliver. At what point do we stop? When is enough enough?

We may find that, on occasions, the decision will be taken for us. For example, it is quite common to be asked to provide a 1 or 2-day course with little regard for what is actually supposed to happen in that time. Thus, resource restrictions may mean that the end point will be known and the question then is what can be achieved in the time?

If our purpose is to produce a proficient workforce, then clearly we need to train to proficiency. Unfortunately, we all differ in the time taken to become proficient at a particular task. In terms of skill development, there is enough evidence to show that training to proficiency is a more effective strategy than, say, unit time or number of attempts. The main problem arising is one of management; it is often too difficult to arrange training systems such that they can handle the open-endedness of training to proficiency. Furthermore, expensive training, such as pilot training, needs a cut-off beyond which any further investment becomes risky. This runs counter to the idea of training to proficiency.

A number of algorithms or decision aids have been produced to assist in deciding on how much training to give on a particular task. One of the most common methods requires training designers to establish, for each task, the criticality of the task, its difficulty and the frequency with which the task is encountered in the workplace – questions included in Table 4.4. The approach is illustrated in Table 6.1. The output from the process is a decision to train to the level of 'awareness', to train to proficiency, to over train or to simply not bother. We need to explain the concept of over-training. We were working in one airline and a line training captain told us that 'it is well known that we forget 30% of what we are taught. So, we teach 130% so that they remember 100%'. If only it were so simple. The flaw in this argument is how do you guarantee that they forget the useless information and not the crucial information. Joking apart, over-training requires additional practice such that the trainees can do the job blindfolded and in their sleep.

Training to the level of awareness requires some treatment of the course content such that trainees know where to look for information and are aware of its existence. The decision not to train is based on similar criteria to those mentioned in Chapter 4. We are not necessarily saying that trainees do not need to know anything about the subject but rather, that the task is so simple or of such little consequence that trainees can be expected to develop the skill with no formal instruction.

Table 6.1
Task Training Requirements

Criticality		High		Medium		Low	
Difficult?		Yes	No	Yes	No	Yes	No
Frequency	High	2	4	2	4	3	4
	Medium	1	3	1	3	3	4
	Low	1	2	1	3	2	4

Key: 1 – Overtrain
     2 – Train to Proficiency
     3 – Train to level of Awareness
     4 – Do not train

Of course, the performance element of the task is only part of the whole course. Skilled performance requires supporting knowledge structures; what we called underpinning knowledge in Chapter 4. More traditional approaches to course design have consigned knowledge to one of three categories; must know, should know and could know. However, in order to meet our goal of delivering effective and efficient training, we cannot afford to waste time delivering instruction that does not support skilled performance. But we still need a method of deciding upon what to include in the course.

Just as in the example of the establishing the overall course aim with which we opened this chapter, we could start by taking each of our training objectives and, through focus groups and brainstorming, map all possible training content onto each objective. Each statement could be then be tested in terms of what would happen if a student did not learn the item.

From our TNA we should have a fairly clear idea of the mental models which support skilled performance and these could be used to verify the need for content and the CTA methodology described in Chapter 4 would also go some way to answering the question. As a result of this exercise, we should end up with an understanding of the depth of training needed to achieve the desired level of skilled performance and an inventory of knowledge items which need to be presented to our trainees. In the next section we will look in more detail at the methods we can use to deliver training.

Learning and Learning Events

In Chapter 3 we looked at learning as a multi-faceted process. First, there were some aspects of the subject matter that we needed to consider. We also saw that learning involved activities external to the learner. These are the events to which learners are exposed and which should promote learning. Finally, there are the internal processes by which information is coded and stored in order to create knowledge. In an ideal world, we would select the optimum activity for each internal process as appropriate for the particular knowledge domain. In the real world, we just have to do our best. The fact remains, though, that we do need to consider how best to present the information to the learner. We are concerned, at this stage, with the overall scheme of delivery, the goals we are trying to accomplish and the medium best suited to achieving our goals.

At a strategic level we can identify four methods of delivery of instruction: Passive Reception, Discovery Learning, Knowledge Deficit and Accrual, Guided Construction. Passive Reception is probably most familiar to the readers of this book. Someone speaks while the class listens. Discovery Learning, on the other hand, involves students in exploring issues through a variety of resources, possibly even setting their own goals and objectives. Knowledge Deficit and Accrual requires us to make some estimation of how much our students already know and then we just fill in the gaps. Finally, Guided Construction can be seen as a form of dialogue between instructor and student during which suggestions and hints are provided which point the student in the right direction. The student is still responsible for filling in the blanks in their knowledge, so to speak. These models, which differ in the extent to which the instructional process is student-directed, represent top-level frameworks within which delivery takes place. We are not talking about the media of instruction, at this stage.

We are trying to identify a suitable, or preferred, course model. Again, organisational constraints will have an impact. For example, Discovery Learning is more time consuming than passive reception. Guided construction is probably more tutor-intensive than some of the other methods.

At a tactical level we can identify some more fine-grained delivery methods: Rote Learning, Direct Instruction, Discovery, Use of Examples, Analogy and, finally, Drill and Practice. Rote Learning involves students learning things 'by heart' such that recall is almost instantaneous. Immediate Action checklists are an example of this. Rote learning does not necessarily involve any degree of understanding on the part of the student. Direct instruction is the more normal activity occurring in most classrooms; instructor-led sessions. Discovery, as discussed above, involves students in exploring a body of material. Much of what adults do in the workplace when trying to discover an answer to a problem can be included under this heading. The use of examples requires students to be exposed to a sufficient number of illustrations of a particular category of object or event. At the time of writing, Controlled Flight into Terrain (CFIT) was considered a significant problem in that it usually resulted in the tragic loss of a serviceable aeroplane, crew and passengers. By looking at all known CFIT accidents over a period of time we hope to learn from the examples offered. Learning from analogies involves examining a system which is representative of the one we are actually studying and looking for insights which will help our understanding. For example, we sometimes say that the human brain is like a computer and much research into cognitive psychology uses the computer as an analogy for the functioning of the brain. Drill and Practice relates more to practical skills and requires the student to rehearse the procedure until it becomes second nature. In Table 6.2 we have tried to show some general relationships between types of learning and optimum learning events.

We have used the term 'delivery method' in the sense of a generic activity that can be used to guide the way information is presented to students. We will see how these delivery methods become packaged into classroom activities later in this chapter. These general activities are all available to us as training designers. How they are used, singly or in combination, is part of the design process. However, we also need to consider another set of requirements. We said earlier that learning occurs on the inside as well as on the outside. The stimulus we present to the learner needs to be somehow transformed internally into a knowledge structure. In the next section we will examine some of the activities at the tactical level that can enhance learning.

Table 6.2.
Learning Outcomes and Learning Events

Declarative
Propositions - Rote
Schema - Direct Instruction
Mental Models - Discovery
Procedural
Rules - Examples
Skills - Analogies
Automatic Skills - Drill and Practice

Events of Instruction

The 'events of instruction' are those activities designed to help trainees to learn efficiently. They function at two levels, macro and micro. On one hand, we can design course units that act as an event whilst, such as a test or a periodic review. On the other hand we can use these activities to guide our design of individual sessions, for example, by including summaries of key points at intervals through a lesson. We are using our knowledge of what seems to promote effective learning, those events external to the learner, to help shape our course design. There are different views of what constitutes a list of instructional events but they can all be grouped into:

- Preparation
- Focusing Attention
- Enhancing Storage
- Assisting Recall
- Promoting Transfer
- Monitoring Performance

Preparation

These activities prepare the learner for what is to follow. At this stage, we establish expectations on the part of the students by giving them an idea of what the course is about, what their part will be and how the course will help them do the job better. At the macro level, our overall course aim will serve this purpose, as will individual lesson objectives at the micro level. We also need to get the students to retrieve any prior knowledge they may have which is relevant to the subject. Recall of prior learning helps to build meaningfulness and is a precursor for generative learning to take place. It is also the first step in building skills that can transfer to other subject areas.

Focusing Attention

At this stage we are presenting the various stimuli, or materials, to the learners and drawing their attention to what is important. We can use visual aids to illustrate significance. We can underline words, use a highlighter, use repetition, even have banners in hallways. At one pilot training school we visited, the flight training rooms had a subtle blue tint to the carpets while the ground-school rooms had a brown hue. At a management training organisation, each of the different syndicate rooms had a different colour scheme which, again, reinforced the difference between the teams and promoted a form of group solidarity. So, we can even paint the walls and change the carpets as a way of focusing attention!

Enhancing Storage

Having presented the information, we need to add ways to ensure that it is effectively stored in memory. Things like summaries, flow charts, diagrams, mnemonics and repetition are all activities which improve retention.

Assisting Recall

Having got the information in there, we need to be able to get it out again. This can only be done by involving the learners in activities that require the learning to be applied. Tests are a method commonly used. We can also get students to complete a task, describe a process, answer a set of 'What if...' questions. In short, anything that gets students doing something is a method of assisting recall. We should also point out at this stage that the act of recalling stored information is a way of checking the effectiveness with which it was stored in the first place. The generation of errors can shed light on the structure of the knowledge stored in memory, flaws in its initial acquisition and also weaknesses in the training method used. The discussion in Chapter 3 between the tutor and student about the weather off the Pacific Coast is an example of using error analysis as a means of checking understanding.

Promoting Transfer

In our earlier discussion of learning as a process, we said that our ultimate goal was to develop expertise. We also said that 'experts' differed from 'novices' in the way they analysed situations and formulated solutions to problems. In short, expertise allows us to generalise across a range of situations. Therefore, we need to give careful consideration to methods we can use to promote the development of transferable skills. We can get students to combine information with prior knowledge, to integrate their learning in order to view the bigger picture, to generate solutions to novel problems on the basis of what they already know. In short, we can work on developing and fine-tuning their mental models.

Monitoring Performance

The final category of instructional activities are those which allow the student to monitor their own progress, the metacognition referred to in Chapter 3. Test results spring to mind as the most common form of performance monitoring. However, we can also give personal feedback that is intended to develop a specific aspect of performance. We can encourage student to self-evaluate. So, for example, we can get students to describe their level of confidence in their own understanding of the course content. The simple question 'What bits of the course are you least happy about?' will elicit evidence of self-monitoring at work. A fuller set of the events of instructions is given in Table 6.3.

As we said earlier, all of these things take place within individual training sessions but we can also see if we need to weave them in to the course as a whole. So, we can set a mid-term exam not just to assess progress but also to focus attention and assist recall. Having identified a set of learning activities, we now need to give consideration to how we can organise the material in order, the sequence in which we will present materials, to promote effective knowledge construction. By building this into our course design we are trying to improve effectiveness of learning from the outset. Having considered possible activities, we will now look at possible models of course sequencing.

Table 6.3
Events of Instruction

Preparation	Select Objectives, Recall Prior Knowledge, Estimate Workload
Focusing Attention	Focus, Search, Contrast with Prior Knowledge, Confirm new Knowledge
Enhancing Storage	Analyse, categorise, outline, paraphrase, use metaphors, examples, elaborate, link to prior knowledge
Assisting Recall	Repetition, Rehearsal, review, Mnemonics, mental rehearsal of skills, utlise
Promoting Transfer	Analyse, synthesise, draw inferences, implications
Monitoring Performance	Test, Judge, Create new problems

Organisation of Content

One of the problems with developing courses is the decision about what order to use in delivering the content. If we look at an existing course, or other courses we have attended, we often find that material is divided into blocks and we often have to study a block until it is complete. It is not unusual to spend a half-day, or even a day, on a single subject. Talking to instructors, we are often told that it is convenient to have this concentration of effort - convenient for the instructors, that is. Whilst one is working, the other instructors can be doing other things. We hear comments to the effect that just having single training sessions free during the day does not allow you to do anything productive and so it is better if one instructor can cover a block of time. Fair comment, possibly, and certainly indicative of the fact that there are several viewpoints which need to be considered. However, if we stick to the purist view for a moment, what models can we use to help us sequence material?

We can adopt one of two distinct themes when considering how best to order our training material; we either deliver instruction in a linear fashion or in a circular fashion. The Linear model requires us to arrange our blocks of instruction in a continuum. We start at the beginning and then work through to the end in accordance with the plan. We may intermix individual lessons in order to bring variety to the day but, essentially, we are treating topics as entities in their own right.

The Circular model requires us to break out groups of objectives which form a hierarchy of stages of training, each of which is of increasing complexity or difficulty. On the first pass, we cover all of the course objectives that fall into the lowest rank in terms of their complexity. On the next pass, we develop each subject slightly and so on, with each succeeding pass becoming more complex.

Applying the Circular Model

As an example of what we mean by the Circular Model of sequencing, we once looked at training Loadmasters on military transport aircraft. Loadmasters are responsible for the management of the cargo bay: loading, tie-down, monitoring in flight and off-loading at the destination. The nature of the operation was such that we could identify layers of operation; flying the aircraft with an empty cargo bay, operations within the UK, flights to military airfields in Northern Europe, flights to North America. Each of these stages involved an increasingly complex set of tasks for the Loadmaster. For example, the paperwork required for Customs officials varied from nil to nightmare along the continuum. The existing method of teaching involved a block of classroom lessons, removed from any operational context, which often proved baffling for students. By dividing the lessons in the manner just described, students were able to master each stage and to build on prior knowledge as their training progressed. Given the nature of the job, most of the course could be segmented this way and, when incorporated into the practical flying exercises, permitted a more efficient approach to training.

An example of the circular model of training is the scheme known as Integrated Job Performance Training.[23] As with the Loadmaster example, the structure of the job is used as a guide for sequencing instruction. Real-world tasks that represent sub-sets of workplace skills are created which act as the focus for a block of instruction, each task becoming more complex until, by the end of the course, students are able to deal with the 'worst-day-at-the-office'. The theoretical elements are layered such that students are building and consolidating their learning as they apply theory to real problems.

Techniques of Sequencing

Having decided on a linear or a circular approach, we still need some means of putting our objectives in order within the individual teaching sessions. For this, we can call on some of the rules of thumb that have been developed over time. For example, training needs to move from the simple to the complex, from the known to the unknown, from the particular to the general and, finally, from the concrete to the abstract.

We can get some experts to help us to cluster training objectives. Write the objectives on a set of Post-it sticky labels and get a focus group to arrange the labels on a flipchart. We can combine and recombine until the objectives have been sorted into a reasonable set of inter-related groups. The resulting set of clusters can help us decide on a sequence of delivery. At the same time, this method will allow us to verify that we have not missed anything, that our set of objectives is complete.

Where we are dealing with practical skills, such as aircraft handling, we can choose one of three approaches to sequencing: Progressive Parts, Cumulative Parts and Backward Chaining. The Progressive Parts approach involves mastering step one before moving on to step two, then step three and so on. The Cumulative Parts model involves practicing step one, then step one and two, then steps one, two and three. Backward Chaining requires the student to achieve mastery in the final element of a skill before then moving backwards a step to do the penultimate step together with the final step. When that is mastered, we move back a further stage in the procedure. So, if we were teaching student pilots to land an aircraft, first we would concentrate on the flare and touch down stage. Once that was accomplished, we would move back to, say, a point at which the aircraft was lined up with the runway, wings level and at the appropriate speed. We would then rehearse the final approach and landing phase until the student was proficient. Then, we would let the student take control at the end of the Down-wind leg at the point at which they need to start the Finals turn. This method has been used to teach US Navy pilots how to land on an aircraft carrier and proved to be more effective, and more efficient, than the conventional approach.[24]

The sequence of instruction will also be influenced by organisational constraints, such as those listed in Table 6.4. However, it is important to remember that we should be sequencing the instruction, as far as possible, to suit the needs of the learners rather than to satisfy the wishes of the staff or some external agency.

Furthermore, as we said earlier, it is quite conceivable that our trainees simply could not reach proficiency in the training time available. It may be that we want to establish an interim graduation performance that will be achieved on the course and then the 'finishing off' will be done once the trainee enters productive service. Thus, part of our planning will include the need for 'on-the-job' training after graduation. One of the reasons for clarifying the course aim at the outset is so that a decision can be made about any alternative strategies that will be needed to cope with training which cannot be covered by the planned course.

Taking Stock

At this stage, it might be worth summarising the discussion so far (an event of instruction, perhaps?). We have said that learning is a process which involves learners being exposed to a set of events which, in turn, results in some internal reaction which we call 'learning'. The internal reaction is the organisation of data to produce knowledge. The knowledge is stored in frameworks and can be recalled at a later stage. Individuals vary in the accuracy and completeness of their internal representation of that knowledge, and also in the ease with which they can retrieve information when required. As training designers we are interested in the nature of the experience which has to be provided for effective learning to occur, both at the macro level of general strategy and at the micro level of events of instruction. Once we have mapped out how we are going to set about organising the material, we need to give consideration to the problem of delivering the training. So what methods are available to us?

Media Selection

Before we answer that question, we just want to make a simple point but one which is often overlooked or missed completely. Most of the stimuli we need to present to learners exist in one of five basic modes:

- The written word - text on a page such as this book, labels on a diagram, etc.
- Spoken word and significant sounds - the narration of a video or audio tape, the engine noise in a simulator, the instructor at the front of the class.
- Still images - photographs, drawings, diagrams, charts, etc.

- Moving images - Video, film, live acting, standing in a hangar and watching the flaps deploy.
- Real objects or representations of objects which can be manipulated.

These are the things students interact within order to gain knowledge. We can combine media in various ways. The method of instruction we use is only a vehicle for exploiting these basic media types. Each method will have strengths and weakness. One combination of media can prove to be more powerful when delivered by method A as opposed to method B. For example, we have stood in a hangar and watched the operation of the undercarriage of an aircraft that had been raised up on a set of jacks and we have seen the same thing on video and as an animation on a computer-based training system. Essentially, it was the same moving image and yet the quality of the experience differed according to a whole set of intangible factors (one of which was the cost involved in setting up the learning experience). We offer this minor diversion as a word of warning. It is possible to set up a technology continuum ranging from a printed book at one end to a full-mission simulator at the other. This differentiation between media and medium of delivery is not always clear in most peoples' minds. All to often we find trainers seduced by the technology - in particular with the advent of multimedia and virtual worlds - with little thought being given to the learning which is being promoted by using that particular medium.

Selection of Training Methods

Having discussed, in very broad terms, the training methods available to us we come to the question of which ones to incorporate in our proposed course. Like much of what has been discussed so far in this book, there is no easy answer. The question can be addressed at a strategic level and it can be addressed at the level of matching stimulus to intended response. At a strategic level we can start by getting answers to the questions in Table 6.4. By mapping the answers to these questions onto the information in Table 6.5, we can start to narrow down the range of possible training methods.

Table 6.4
Factors Affecting Choice of Training Media

Number of students on each course.
Type of Learner (entry level, learning style)
Type of Course Content
Lead Time for Course Production
Frequency of Course
Duration of Course
Location of Students
Availability of Staff (teaching and support)
Funds Available
Space Available
Attitude of Organisation

With that in mind, we can now look at the various methods available which will allow us to package media in order to achieve our various events of instruction. This list contains most of the training methods in common use:

Large Group
    Lecture
    Video/ TV

Small Group
    Lesson
    Discussion
    Case-study
    Role-play
    Simulation

Individual
    Individualised Learning (text-base materials, CBT)
    Task Rehearsal (repetition)

We have assumed that most readers of this book will be familiar with the various methods. It is not our intention to go into the production of individual sessions within each of these categories; we simply do not have the space. Our concern is more with the process of selecting a method best suited to achieving our training goal. The Table 6.5 lists some of the main advantages and disadvantages for each group of methods.

However, given our contention that some methods are more suited to particular events of instruction, we need a more sensitive tool to aid our choice. In Table 6.6 we have attempted to map training methods onto learning outcomes. Of course, the choice of method at the micro level is equally bounded by the organisational constraints listed in Table 6.4. However, the suggestions in Table 6.6 should be considered a starting point for subsequent negotiation.

Table 6.5
Advantages and Disadvantages of Training Methods

Large Group
    Advantages
        Allows bulk processing
        Can be motivational if presenter is interesting
        Can be rapidly updated
        Cost-effective
        Minimal logistical problems
        Content and pace easily controlled
        Best for facts and procedures
    Disadvantages
        Passive
        Takes no account of individual differences between learners
        Unsuitable for some topics, such as skills development

Small Group
    Advantages
        More active involvement of participants
        Able to address higher order skills and attitudes
    Disadvantages
        Less cost-effective
        Time-consuming

Table 6.5 Continued

Individual
    Advantages
        Flexible
        Addresses individual needs
        Increased learner control
        Good for facts, procedures, concepts and principles
    Disadvantages
        Requires high levels of self-motivation
        Materials can be difficult to develop
        Logistical constraints
        Weak for social skills development

By now, we should have mapped our TNA onto a set of training activities. We will have considered our broad strategy and have identified appropriate events that best suit the type of learning we want to achieve. We should have considered the sequence in which we want to deliver the training and the depth of instruction needed. We should also have identified a range of delivery methods. We can now start to allocate tasks to the developers of the individual lessons. We can prepare a lesson specification that sets out the relevant objectives, position of the lesson within a sequence (and, thus, the student's prior knowledge), a rough time budget and a guide to the preferred method of delivery. All we have to do now is sit back and wait!

Table 6.6
Teaching Methods and Learning Outcomes

	Lecture	TV Video	Lesson	Discu-ssion	Case-Study	Role Play	Simul-ation	Self Study
Proposition	✓		✓					✓
Schema	✓	✓	✓		✓			✓
Mental Model		✓		✓	✓	✓		
Rules	✓	✓	✓	✓		✓	✓	✓
Skills						✓	✓	

Conclusion

In this chapter we have tried to sketch out some of the issues around the selection of training events. We have been concerned with just what we are going to do with the trainees once we have them in our grasp. We have looked at the need for a clear overall goal for the course, not just to keep the course focussed but to keep the design process on track. We have considered ways of sequencing material. We developed our earlier discussion of learning as a process to see what the implications were for the selection of activities that we could include in our course. We have spent some time discussing how to package the training and we ended by considering at what point we stop training.

By now we should have a fairly clear idea of what our course is going to look like. We have scoped out the organisational issues that we will need to address and have confirmed a set of appropriate objectives. We have now identified a set of activities that will allow students to achieve the objectives and have arranged them into a sequence that will promote maximal learning. We have even identified the most appropriate media for each element. All we need to do now is create the courseware.

Case Study

You have noticed an increasing trend in violent behaviour directed towards cabin crew as well as some of the ground passenger handling agents. You have been called in to speak to the Director of Cabin Services who suggests that it is time to put together some training to prepare staff for dealing with unruly passengers. On further investigation, you find that training staff to restrain passengers could have legal complications and so you decide to limit the objectives to detecting the warning signs of a problem brewing and techniques for calming things down. You have been told that all personnel must be competent in dealing with problem situations, so this solves the problem of your graduation standard - or does it? Anyway, you sit down to plan the course. What factors are you going to consider in terms of the delivery of the training?

7 Measuring the Effectiveness of Training

Introduction

Now that we have designed our course, we need to consider if it works - or not. In this chapter we will consider the methods by which we attempt to evaluate the effectiveness of our training. You might be forgiven for thinking that we need to actually run the course first before we can say whether it is successful or not. However, at the planning stage, it is worth spending some time thinking about the measures of performance you will need to have in place if you want a convincing answer to the question of effectiveness. In general terms we are interested in the concept of validation - the extent to which training meets declared goals - and evaluation - the extent to which transfer occurs to the workplace. By the end of this chapter you will have:

- Considered the need for the measurement of effectiveness.
- Examined methods of internal course validation.
- Examined methods of evaluating training transfer.
- Considered the broader implications of change within the organisation.

Motives for Measuring

There are many reasons why we might want to try to assess how well our course works. First on the list, and probably the most obvious, is that of simply being interested in whether training interventions do what they are supposed to do. We all know people who have failed a training course and, certainly in the UK, many families have their summer holiday in the shadow of the impending annual announcement of school exam results. However, was that poor performance on the course or those disappointing exam results the fault of the individual or the course they were given? In the field of CRM training there is a group of pilots, the 'boomerangs' or 'hard-nails', for whom CRM has had zero or even a negative effect. Ignoring any discussion about personality and attitude, the fact remains that the training event did not achieve its goal. Some flight simulators have been found to cause decay in aircraft handling skills rather than an augmentation. Negative training effects are very real and can only be detected through evaluation.

We also need to consider the impact of training on students' attitudes. Many readers of this book will identify with the situation where a student walks out of the training room thinking 'that was a waste of time'! Evaluating training events gives us the data to ensure that, at the least, our customers leave with a positive motivation to come back next time. It will also enable us to better tailor training to development needs.

Mention of the word customer highlights the increasingly commercial relationship between training provider and user. Total Quality models establish clear conceptual relationships in these terms, but as soon as money changes hands, even the notional charges used in some internal accounting system, the relationship becomes real. Now training departments have to compete for custom, even from staff who work for the same parent company. Therefore we want to make sure that our product offers the best value for money. How can we win that trade if we cannot show proof that our training is best? Evaluation cannot give cast-iron guarantees, but what conclusions can be drawn from no data at all?

Product liability suits are a growth industry for the legal profession. Training is the time when, as far as possible, the risk of a trainee making a mistake is reduced to a minimum before entering productive service. However, because of the gap between training and the real world we induce a secondary risk; we cannot predict what mistakes trainees will make based on their incorrect interpretation of and extrapolation from the course content. We need only to think back to the pilot whose control system froze up in Chapter 3 to appreciate the problem. Without wishing to seem alarmist, it does seem that, in the event of an accident, questions could be asked about the link between training effectiveness and graduate performance. Unless we can prove that we have done our best to make sure that no latent errors are transmitted through our training, it seems that, in today's increasingly litigious society, we need to be ready for that the summons to appear in court to land on our doormat.

In Chapter 2 we considered the range of stakeholders who have an interest in training, one of whom was the regulatory body. In most countries, the pilot ground examination results offer a vehicle for keeping track of the performance of the school delivering the training. The perceived poor quality of CRM training being offered in the UK led to the Civil Aviation Authority investing in the definition of a set of instructor performance standards. These standards are, at the time of writing, being considered for inclusion in the Joint European CRM training requirements. Thus, a crude form of evaluation (Regulatory dissatisfaction) has resulted in action on the part of a stakeholder.

We can summarise the process of evaluation in terms of who is interested and what purposes does it serve. In terms of who is interested, the key groups of clients are the delegates on the course, those who deliver training, those who design training and, finally, those who authorise or pay for the training. The purposes of evaluation can be categorised thus: should we continue to do what we are doing, which is an attempt to prove that our solution works, or should we be doing things differently? This second purpose can be further clarified. On the one hand, we can ask questions of our approach that could lead to refinements and improvements or, on the other hand, we can use evaluation techniques to compare alternative approaches to training; should we use method A or method B? In that vein, we will look at case study later in this chapter that was triggered by a need to justify an investment in an expensive CBT system to replace classroom instruction. We can add a third purpose that considers evaluation to be an activity of learning. By forcing course members to reflect upon the experience, we can assist in the consolidation of their learning.

Validation and Evaluation

We can pull together these ideas about reasons for measuring effectiveness by looking at the stages in the training process at which we can intervene to make measurements. We can make a distinction between student performance on the course and the subsequent application of the course to the workplace and, earlier, we gave these stages the titles validation and evaluation. Both are attempts to measure effectiveness. In the case of validation we are trying to see if the course we have designed allows students to achieve the objectives we have identified for that course. In effect, have we achieved what we set out to achieve? We can qualify the question by adding 'by the most efficient and cost-effective means'. It presupposes that we selected the correct aim and set of supporting training objectives in the first place. It also assumes that we selected the right cut-off point for terminating training. What we now need to do is check that our methods were correct.

Once our students leave the training room, they need to reach a level of productive service as rapidly as possible. Evaluation is intended to answer the question 'has our training allowed our graduates to become productive'. Again, we can qualify the question; have they reached the standard in the shortest possible time? Were our methods the most cost-effective? Is there any aspect of the task that we have missed? The broad strategies available for the measurement of effectiveness[25] are listed in Table 7.1.

Table 7.1
Evaluation Strategies

Level 1   Reaction - did the students enjoy themselves?
Level 2   Learning - did the students achieve the course objectives?
Level 3   Behaviour - can students do the job?
Level 4   Results - what benefit do trained students bring to the organisation?
Level 5   Benefits - how does society as a whole benefit from the training?

Levels 1 and 2 can be seen as aspects of Validation, while Levels 3 – 5 are what we would consider Evaluation. We can see immediately that some of the dimensions of effectiveness of training are a sub-set of the quality management task and, therefore, might be the responsibility of an airline's quality audit team. We deal with quality management in more detail in Chapter 9. However, we will treat validation and evaluation in detail in this chapter on the grounds that effective implementation of a measurement regime requires prior planning and, so, can reasonably be considered the responsibility of the original training designers.

Some General Observations on Measurement

Having outlined the reasons why you might want to measure performance and identified those stages where you can go looking at data, we now need to look at some measurement techniques. Before we do that we want to look at some methodological problems. Evaluation requires us to look for data which will demonstrate that what we think is occurring in training is actually taking place. The aim of this book is to support trainers involved in course design and we have already said that we need to think about an evaluation strategy as part of the design process. However, it may be that we have been tasked with updating an existing course. We also said earlier in this chapter that evaluation allows us to compare different training systems. We may need to devise a strategy that will allow us to see if any changes we make to an existing course are having the desired effect. The observations in this section apply equally to the range of situations we have just outlined.

Any assessment of effectiveness needs a starting point. How can we tell if things have changed if we do not know how well we were doing before we spent the money on, say, new textbooks? Similarly, how do we know we are teaching the right things if we have no learning goals? We once got 12 months into a CBT project before the client even understood why learning objectives were needed. They simply wanted to automate what happened in the classroom. The ground instructors had no objectives; they adapted their lessons to what they thought the students needed to hear. In effect, no two courses ever got taught the same information. What confidence can a customer of that school have in its workforce if its employees have all been taught something different? Before we can consider measuring effectiveness, we need to know what the system is trying to do – we need objectives or some statement of desired outcomes. It is then our job to find out how close the course comes to achieving its aim.

The data we use in coming to our conclusions can be objective or subjective. Objective data includes examination scores or certain task accomplishments in the simulator. Subjective data includes students' views about the course, managers views about graduates once they enter service and even instructors feelings about whether they are getting through to students. It is probably appropriate at this stage to consider the difference between evidence and proof. We once ran a series of CRM courses for all of the pilots of a small airline. On subsequent visits to the airline, First Officers (FOs) told us that they had noticed a change in the behaviour of some of the more difficult Captains after the course. One FO observed that a particular Captain, famous for his short fuse and violent temper, could now be seen visibly trying to control himself whereas, prior to the course, he would simply have let rip. The observations of the FOs can be considered evidence that something happened but could not be considered proof that our training course was effective in the way we intended it to be. It is probably fair to say that evidence, as opposed to proof, is the most likely outcome of any attempt to measure effectiveness. However, such results should not be rejected because they cannot be subjected to a statistical test. In some circumstances we can seek proof, such as competence in a particular task but, more often than not, we will need to be content with evidence.

Data can be gathered directly, by asking questions of students or by looking at various outputs such as exam scores or traces from plotting devices, or indirectly through questionnaires or through the views of third parties. Data can also be gathered contemporaneously, during an exercise or before a course has ended, or after a time lag.

Data will also require processing. We once ran a course for just over 100 pilots. The course comprised 17 sessions, each of which was evaluated against five criteria. Thus, for each course member, assuming they filled in the feedback form completely, we collected 85 bits of data. Multiply this by the total number of students and you can see that we had quite a data processing task on our hands. We will return to this topic later.

In summary, all studies of effectiveness, if they are to be of any real benefit to training managers, need a dimension to be studied, criteria for success and practical methods of sampling. Because of the nature of the process, the same methods can be applied across a range of situations; you do not have to reinvent the wheel and borrowing of other people's evaluation techniques is a sensible idea. If change is the object of the study, then measurement needs to be taken before and after. It is probably fair to say that a significant number of effectiveness studies are rendered invalid because of the failure to measure the base state prior to the intervention. We will consider some of these aspects in more detail later in the chapter. For now, we will look at measures of effectiveness that can be used while students are in training.

Internal Measures of Effectiveness

Analysis at Level 1 - satisfaction - is probably the most common form of internal measurement of effectiveness made whilst students are under training and, in fact, research in the UK revealed that few companies bother to go beyond this stage. Colloquially known as 'Happy Sheets', Level 1 investigations typically ask the students if they found the course content interesting and if they thought that the information was useful or important for the job. In a study of happy sheets collected after pilot CRM training in a large airline, responses fell into three groups. The first group marked all sessions highly on both dimensions while the second, much smaller, group gave everything poor marks. As an aside, in an airline we worked with, we found that 50% of the below-average assessments came from 10% of the delegates.

The third group in the Study was probably the most useful. This group was able to discriminate more effectively between sessions, giving a spread of scores that differed between sessions and between dimensions. In the first two groups, the scores on both dimensions were very strongly correlated; either all high or all low. In the third group, respondents seem to have given more thought to their responses.

At Level 1 we are measuring attitudes; what do people think about our course. In reality we are probably more interested in performance. We want to establish a relationship between the training delivered and the achievement of the stated objectives. We can ask students if they thought the course helped them to achieve the graduation standard. We can ask them if any subjects were treated in too much detail, in insufficient detail or were simply missing altogether. We need to remember, however, that for introductory courses, few trainees will have experience of the actual task and so are not in a position to comment on the effectiveness of the training in preparing them for the job. Hence, questions about depth of coverage are likely to elicit responses that reflect little more than the extent to which we managed to sustain the students' interest.

Level 2 evaluation is largely covered by the various testing methods used during our course. Although Level 1 measures collect opinions, they can still provide some useful information which we can use to improve the course. Unfortunately, opinions are probably not the best evidence at Level 2. For this we need to turn to the output from exams and performance tests. The weaknesses of the various forms of testing were discussed in Chapter 5 and apply equally here. Just as 'happy sheets' can give a false confidence in our course design, so good exam results can be equally misleading.

External Methods of Measurement

At Level 3 we are trying to measure the extent to which training transfers to the real world. After all, if training does not produce employable staff, then something has gone wrong. We can apply a number of measures at this stage. We can see how much of the job a trainee can undertake without supervision and without making a mistake. We can measure how much supervision trainees need, usually measured in time, until they can be considered competent. We can look for gaps in knowledge. We can look at reported errors and see if they correlate with time in employment.

Because training is never simple, we need to add a word of caution at this stage. Evidence that is gathered at Level 3 might not be indicative of a clear cause and effect relationship. Poor performance in the workplace may be the result of inadequate training, but it can also stem from a variety of other factors. For example, on one follow up study we did, graduates from a supervisory management training course told us that, on their return to work, they were told by their foreman that he did not care what they had been taught on the course. Whilst he was around, they would do things his way. In this case the organisational climate prevented the application of skills learnt in training.

At Level 4 we are looking for a real return on the investment we have made in the individual. At this stage we are not simply concerned with their ability to do the job. We are looking for creativity in problem solving, above-average performance, enhanced commitment. In short, we are seeking added-value. The first problem we will encounter is that of selecting a reliable performance indicator. For example, an engineering company was concerned that the university graduates joining its fast-track management programme never stayed long in the company. Its solution was to develop an enhanced induction training course. The effect on retention was instantaneous. But how did they establish the link between the training course and graduate retention? In this case, they made sure that all other variables, such as the selection process, conditions of employment etc, remained constant.

Finally, at Level 5 we are interested in the broader effects of our training. At this level we are interested in seeing if training is having any broader societal benefits. Thus, we may choose to establish minimum standards of literacy and numeracy for our workforce and implement training courses for those that fail to meet the standard. We might then be interested to see if the training has any effect on the lives of our employees outside of work. Again, we might be interested to see if safety awareness training reduces the rate of domestic accidents.

In Level 4 and 5 evaluations, we need to allow time for the training effect to work through to performance in the workplace. Therefore, any attempt to evaluate needs to planned for some time after graduation. We now encounter a new problem. If we wait, say, six months before we evaluate, to what extent are we measuring the effects of the training course and to what extent is the observed performance the result of expertise acquired through contact with skilled operators? We can also see that, as we move up the evaluation levels, the problem of finding the evidence becomes increasingly difficult and, consequently, costly. Therefore, it may be hard to justify, in financial terms, the exploitation of all levels of evaluation. This is particularly the case at Level 5 as the benefits are likely to be perceived as accruing more to the world outside of the company.

Sampling

Having decided on a measurement strategy, we need to give some further thought to data collection. We mentioned earlier that it is possible to become swamped by the data. Some organisations routinely collect Level 1 data from every course. By using machine-readable forms the processing task is made easier. Charts are produced which show trends over time. Unfortunately, the variation of the average score for each course is rarely more than a couple of percentage points and is certainly not significant. There is danger that we end up collecting data for data's sake. As part of our strategy, it seems to make sense to use performance measurement as a way of answering specific questions. We can sample for a period once the course has been introduced to make sure that it works. We can repeat after an interval to check that things are still working. We can sample after changes, either to the course – such as the introduction of a new instructor – or to the entry standard of the trainees. In short, we suggest that evaluation be used as a management tool to meet a clearly-defined objective rather than as a security blanket to make us feel that we are doing the right thing.

Next, we need to think about sample sizes. How many users must we question for the results to be significant? Most of us know that questionnaires and postal surveys are notorious for poor return rates. We are also familiar with the concept of a minimum number of samples from a population being needed for a conclusion to be a valid for the total population. But evaluation is all about looking for indicators of effectiveness. We want to see where shortfalls occur, not establish a statistically valid picture of the student population. In the next section we will look at a case study which used questionnaires as one of the measurement tools. The questionnaires were sent to a potential pool of 43 ex-students and 11 responses received. At nearly 25%, return rate was quite respectable. However, after the first eight responses, nothing new was found in the next three returns. The confidence levels that the conclusions were sound were building all the time but in terms of identifying what problems with the course design needed rectification, the sample was adequate. Remember, we said earlier that, in evaluation, we sometimes have to deal with evidence in the absence of proof.

Case Study - An Evaluation of a Pilot Aircraft Type Conversion Course

As an example of some of the pitfalls that can be encountered during an attempt to measure training effectiveness, we just want to describe an evaluation project we undertook on behalf of a company that provides initial and recurrent flight simulator training on behalf of other airlines. As part of their 10 day course to train new pilots how to operate a particular aircraft type, they had recently introduced a CBT package to replace the classroom instruction. We were asked to see if the CBT was as effective as the previous classroom instruction and, more importantly, whether it was worth the money spent on it.

The course was based around a series of CBT sessions that were reinforced by exercises in the simulator. Once the aircraft systems had been mastered, pilots then concentrated on aircraft handling in the simulator. On successfully passing the course, the pilots then returned to their airline to start flying the real aircraft. The approach adopted for the study was based on a number of assumptions. We postulated that learning would happen in two stages. The first stage involved putting knowledge gained from the CBT into practice in the simulator. The next stage came when students moved onto the real aircraft. Furthermore, students would be able to identify where they felt that there were gaps in training once they started to put the learning into effect. In this, we were trying to exploit students' awareness of their own learning needs in order to identify shortfalls in the CBT.

Having said that the primary aim of the evaluation was to see if the CBT had worked, the first problem we encountered was that no benchmark existed from before the change in training. From the outset we were unable to answer the first question we had been set which dealt with the comparison between the two approaches. In effect, we were now establishing the benchmark of training performance and seeing if there were any gaps in provision.

The only objective data available came from 40 students who sat the five Phase Tests set during the course and the End of Course test. These pencil-and-paper tests represented a measure of accomplishment in terms of mastering the technical complexity of the aircraft. Subjective data came from students during training and graduates who were now flying the line. Students were given questionnaires and asked to rate each of the training interventions (CBT, instructor, fixed-base trainer and simulator) on a four-point scale. They were also asked to rate the individual CBT modules and to offer comments as appropriate. The questionnaire simply aimed to identify gaps in training coverage.

The categories used on the rating scale (Good, Acceptable, Minor Objections and Major Concerns) were sufficiently intuitive to identify strength of reaction but did not call for fine-grained analysis of training. Students were also asked to complete some open-ended questions that were designed to identify specific weaknesses in training coverage. We found that there was good correlation between ratings of instructional elements and the comments in the open-ended questions. Thus, modules that attracted 'Major Concern' ratings were also likely to be the subject of comments.

Given the two-stage model of learning, any significant gaps in the classroom instruction would be spotted by students once they got into the simulator. Gaps in the whole course would become apparent once graduates started flying the real aircraft.

Problems with the objective data, the exam scores, were discussed in detail in Chapter 5 when we looked at test validity and reliability. The high pass rates were found to be of little value in determining the effectiveness of the CBT, which meant that we were dependent upon the questionnaires as the main source of data.

We made the assumption that the students in training would have a clear recall of the content of the CBT. For the course graduates now in employment we would have to accept that their memories of the course might have faded, even though the follow-up questionnaire was administered within 6 months of finishing the course. As it happened, one student stated that he could not remember anything about one of the CBT modules and one graduate said the same about the whole of the CBT.

We also found that some respondents chose not to answer the questions we had asked but, instead, give a general blanket response. Others gave their general opinions about training rather than about this course in particular.

So, did the evaluation work? The requirement to demonstrate that the CBT investment was sound was not fulfilled for 2 reasons. First, exam results (which, we have already said, were flawed) were not available for students who had only used the CBT. Second, we did not have access to students from the previous system who could have given us information about the relationship of the course to line flying. We were unable, therefore, to comment of the comparative benefits of investing in technology-based training.

The next question we can consider is does the current course work? From the evaluation, we were able to identify flaws in the design of the CBT but we doubt if we could separate out the learning effect of the CBT from the other factors involved, such as the students' study of the aircraft manuals outside of class.

With regards to what we could be doing better, the responses from the graduates flying the line did offer conclusive evidence of where the existing provision was inadequate.

This case study shows some of the problems that can be encountered when trying to evaluate training. Meaningful evaluations can be achieved with relatively small sample sizes. However, the methods of collecting the samples require some thought. For example, we were surprised at the extent to which students were unable to recall training delivered just a few days prior to the investigation. As for lessons learnt, we should have made more use of interviews. As it happened, there were constraints on us that prevented talking to any of our respondents. We found that the questionnaire approach can give rough categories of problem but these should be followed up by discussion. Again, rather than try to evaluate whole courses, it is probably better to investigate areas of interest or concern.

What to do with the Results

Having conducted the evaluation, what do you do with the results? We mentioned earlier the problem of simply collecting data for data's sake. However, in one flying training organisation we visited, exam results were plotted on a chart for a six month period and then discarded when the next six month period started. Although short-term blips were detectable using this system, longer-term trends were not.

Stakeholders will have differing interests in the output from the evaluation process. Returning to the key stakeholders mentioned at the start of this chapter, trainees will appreciate the chance to make an input to the process although most are probably more keen to get away from the classroom as fast as possible. We mentioned earlier, though, how evaluation could be seen as an event of instruction. In one study, students observed that the end-of-course evaluation was rushed and that they would have appreciated more time to gather their thoughts. The implication being that a carefully-designed form which forces trainees to reflect on their experiences will probably provide more valuable data as well as consolidating student's learning. However, most evaluation happens after the average student has left the training course far behind them.

The deliverers of training are keen to iron out any aspects of the course that cause difficulties during daily operations. Training designers are more interested in testing the design concept. Finally, budget holders want value for money.

Whatever our interest, it is fair to say that evaluation of training can lead to discomfort. We have all sat through an end of course wash up and every time we raise a point the staff give us a reason why it couldn't work that way. Or they say that that's how they used to do it and they only changed it to this way because of what other students said on previous debriefs. If they already know all the answers, why not just let us get home earlier?

Evaluation is often seen as criticism and therefore provokes a defensive reaction. In reality it is a process which identifies improvement opportunities and, as such, links well with the concepts of Total Quality Management which will be explored in Chapter 9. On receipt of an evaluation, the appropriate response might be to take no action. It might be just to investigate further. What we cannot afford to do is to shoot the messenger.

Conclusion

Measuring effectiveness can be done for many reasons. For example, to make sure that our training actually enables students to do the job in the real world or to see if investment in a new training device makes any difference to the effectiveness of the course.

Evaluation need not involve vast numbers of people. Questionnaires can be used to identify target areas that can be followed up with interviews. Small groups still give valid information.

If you are going to measure, then be prepared to do something with the results. We were once asked to set up a scheme for monitoring the classroom performance of instructors. As part of the planning we asked what money had been set aside to deal with issues such as inadequate visual aids and what plans were there for additional training for instructors; two outcomes which may well have arisen from the process. The project was not implemented because no money was likely to be available nor was there any intention to implement in-house training. To evaluate with no thought for what is likely to happen as a result of that activity is probably more damaging than not evaluating at all.

Case Study

As a result of the harmonisation of regulations across Europe, your airline has had to change some of the procedures for calculating aircraft performance data. This will require an amendment to the aircraft manual be published. The Chief Pilot reckons that a simple letter advising pilots of the change and telling them to read the amendment will suffice. As the Training Manager, you are not convinced. You feel the need for something a little more positive, considering the dangers associated with getting the calculations wrong. As a result, you decide to propose a trial of the planned approach and your own solution, which is a brief self-study guide. In order to convince the Chief Pilot that your solution should be made available to the pilots, you want to conduct an evaluation of the two approaches. Plan your evaluation strategy.

8 Project Management

Introduction

Training design is an activity that requires a number of tasks to be accomplished over a period of time. It can involve more than one person, will consume resources and may require some planning and forethought. It can also be critical in that the consequence of not being able to train on time could have financial or safety implications. In the case of the Boeing 747 landing with the nose-wheel up given in Chapter 1, poor project management was cited as a causal factor at the subsequent Inquiry. We will look more closely at this incident in the next section. It is fair to say, therefore, that some project management skills might be useful in even the simplest of training design projects.   This chapter will look at the management activities associated with running training design projects. By the end of the chapter you will have:

- Considered the stages of the project management cycle.
- Identified the costs associated with training design activity.
- Established the risks associated with training design projects.
- Considered some tools to simplify project management.
- Examined some issues associated with the management of change.

A Project Management Case Study

The initial decision to open a new route and to introduce a new aircraft type, the Boeing 747, was taken in March 1994 although the idea had been around since late the previous year. A project manager was appointed but no formal team was set up. Instead, individuals were appointed in each department. The project manager did not involve himself in the individual departments and control was exercised through regular meetings with individual department representatives. No project management tools were used although macro-milestones were established. The Managing Director (MD) met with the project manager at intervals. Then, in July 1994 the project manager left the company and the MD assumed control of the project.

In March 1994 the project manager started looking for a provider of training for the flight deck crews. A training plan was created which represented the first three flight deck courses. The initial idea was to subcontract all flight deck training to another airline already experienced in B-747 operations and using that airline's aircraft. An airline was chosen but the sub-contractor's pilots' union would only agree to do the ground school and flight simulator instruction. They would not agree to conduct the line training element, that is, the training done on the aircraft. The contingency plan involved a second airline doing the line training. It was also assumed that a limited number of contract pilots and flight engineers would be used from the inauguration of the service in September 1994 through to January 1995 in order to get the service up and running. As part of the contingency plan, these contract crews would be used for line training and checking of the company's pilots.

In May 1994 an announcement was made that the company would take its aircraft from Singapore Airlines. As part of the initial negotiations, the airline providing the initial training had assumed that it would also be providing the aircraft which would eventually be used in service. When this news broke, they withdrew from the whole deal. The start date for training now slipped from May to June which meant that line training could not be completed in time for the first revenue flight. It was decided that the contract pilots would do the line training for the company pilots once the revenue flights had started. The delay also meant that no training would be available for the contract crews.

At the end of May the MDs of both the airline and the initial training sub-contractor met to resolve the dispute which resulted in the offer of ground and simulator training being reinstated. Because of the delay, it was suggested that the training be conducted in accordance with the sub-contractors methods and operational philosophies. Finally it was agreed that additional training would be given to the airline to cover any differences between the two companies' ways of doing things.

Training started on 20 June for the first course, followed by a second course on 4 July. Immediately problems surfaced as a result of the two companies' different operating philosophies. Basic documentation had not been prepared and items were omitted from checklists. Modifications made between the first and second courses were described as 'disorganised, unnecessary and confusing'.

A Project Management Cycle

Typically, a project will go through a number of stages and you may want to the review case study summarised above in the light of the following discussion. Not all training projects are as complex as the introduction of a new aircraft type. Much training design activity in airlines is at the lower end of the scale, involving individuals or pairs of trainers assigned to a task. That said, the stages listed in Table 8.1 can usually be recognised in projects of any size.

Table 8.1
Stages of a Project

Initiation
Budgeting
Scheduling
Resource Allocation
Monitoring
Project Control
Termination

Each stage has its own particular problems and each will have an impact on the overall progress of the project. As part of our own training design activity, we need to be mindful of the management aspects and to take the appropriate action. In the next few paragraphs we will look at each of the stages in turn and consider the implications for training design.

Initiation

The initiation phase of a project requires us to establish exactly what we want to achieve. The identification of a course goal, as discussed in Chapter 6, and the definition of supporting course objectives, covered in Chapter 5, form a part of that process. For evaluation projects, we need to have a clear idea of the question we want to answer. Without these, we are in danger of wasting time and effort. We have already referred to training design projects that started with no training objectives identified and, worse, no understanding on the part of the client of the need for objectives. But at the initiation stage we need to do more than simply state the course aim.

As well as offering a proposed purpose for running the course, subject to clarification once we get into training analysis proper, we need to provide an outline of the project, identifying in very broad terms our requirements for resources and, finally, establishing approximate time scales. We might also want to identify any significant constraints that may apply. We need to establish levels of accountability and authority. The output from our stakeholder analysis, discussed in Chapter 2, will be of use in identifying responsible agencies. It is worthwhile drafting a brief synopsis of the project at this stage which can be used for management briefing purposes and as part of the project control process.

Budgeting

We have already mentioned that training costs money, even if designed and delivered in-house. One airline we worked with wanted a 2-hour team-building exercise to be included in their annual safety refresher training. To create that exercise, a captain and two FOs had been allocated the task. It was estimated that the job would be the equivalent of a person/year of work. We will consider production times for course materials later but the resource allocated to the project in question amounted to something in the region of £80,000 in salaries alone. An outside consultant would have looked to do the job in one fiftieth of the time and at one twentieth of the cost. However, because no cash changed hands, the training was deemed to be at no cost to the company.

As part of our project management, it will do no harm to take stock of the costs involved. Like the airline just mentioned, we may have no intention of actually putting the work out to contract. However, in an increasingly cost-conscious world, training has to be seen to contribute to the bottom-line. Therefore, accurate costs can be used as part of our training decision-making.

We can draw up a list of costs likely to be associated with the design and delivery of training. A representative list is given in Table 8.2. In terms of the training design cycle, a representative breakdown of costs would be:

- Analysis   20%
- Development   35%
- Delivery   35%
- Evaluation   10%

Of course, the actual distribution of costs will vary, the main differences being in Development and Delivery. If we are simply updating an existing course then Development costs will be proportionately less. If we are rolling out a training programme across a large workforce, then Delivery will constitute a higher proportion of the total.

Table 8.2
Representative List of Training Costs

Salaries of Training Staff
Salaries of Trainees
Travel and Subsistence
Office Supplies
Course Materials
Printing and Photocopying
Outside Services
Equipment Expenses
Maintenance Costs
Registration Fees
Use of Facilities
General Overheads
Miscellaneous

Our interest, at this stage, is to identify any investment that will be needed for our course and to give sufficient warning to the Finance

Department such that the necessary action can be taken. The construction of a notional budget will also permit a form of project control.

Scheduling

The scheduling section of our project plan deals with making resources available when required. We know that, once the course is ready, we will need Instructors and Students. We will also need training accommodation and, perhaps, specialist training devices. Where equipment is being procured, we need to consider lead times for production. At the very least, there is minimum time the Rostering Office needs for nominating staff and students who may be allocated to flying duties.

Our scheduling needs to take account of the time needed to prepare training materials. It is impossible to be precise in specifying course production times but, as a rule of thumb, 10-20 hours of production time is needed for each hour in the classroom. If we want to produce classroom materials which can be given to other instructors to be used as a turn-key solution, the additional supporting and explanatory materials required will push production times up by a factor of five. Self-study courses can take 80-120 hours of production per hour of student study and figures of 350-500 to 1 have been quoted for CBT. These figures can be applied to the planned course length, factored for any specific teaching methods, and included in the schedule. We need to be wary of using the lowest figure in each range; undoubtedly something will go wrong at some point and we need to allow a buffer in our planning.

As well as production time, we need to allow for any requirement to seek approval and certification of the course. We may need to negotiate access to copyrighted materials. We certainly will need to pre-test the course and allow time to make modifications. Our experience shows that it takes three iterations for a course to bed down, at which stage it is wise to consider making changes. Of course, if our training design has been faulty, we may have a serious problem on our hands. Take this as a salutary lesson. As the need for CRM training extended to cabin crew, many airlines gave the task to their pilot CRM trainers. Quite often, the pilots took their existing two-day course and collapsed it into a single day event. One airline we worked with had based a significant part of the course around segments of the movie 'Apollo 13'. The appeal of the movie for pilots is readily apparent. However, when the course was first run, the audience reaction to the movie-based exercise was contrary to the trainers' expectations. Cabin crew interpretations of events were at odds with those of the pilots. The course failed and had to be completely re-designed.

One key lesson that we can draw from this experience is that it is important to know your audience. In order to gather information, we need to do research; that's what the analysis phase is all about. However, research requires access to people. In most companies it is difficult to gain access to people during the working day. Airlines have the added problem of Crew Duty Time limitations and Rosters. If we need access to a particular expert, then the problem can be magnified.

In some cases, airlines will get around this problem by using office-based staff as interviewees. For example, Subject Matter Experts are often provided from pilots who have been given management responsibility and who are available during their planned office days. Often we need to make use of cabin crew who are in training roles and who have reduced flying commitments or who, if female, are pregnant and are assigned ground-based duties prior to maternity leave. The problem with this sample is that they are often above-average employees and not necessarily representative of the workforce as a whole. Scheduling is probably one of the more important aspects of project management in training design. It is ignored at your peril.

Resource Allocation

It is probably fair to say that most of our training design projects are going to involve minimal resources and large-scale projects will be in competition for resources with other elements. The scheduling activity will have identified at what stage the work needs to be done. We should have a clear understanding of where our resources are. However, resource allocation needs to be considered over the whole life cycle of the project. For example, what provision is made for general housekeeping and up-dating materials? If we consider the case of recurrent CRM training in Europe, many airlines have found it hard enough to develop the basic introductory course in-house. The regulations require recurrent training to start 12 months after the initial training. The resources needed to start the next design phase whilst the current course is still being delivered are often beyond capability of many airlines.

And we do not need to remind you that most of you who will be allocated to training design will not have given up the day job – training design and flying the line do not mix.

Monitoring

We have talked already of groups such as management, course sponsors and budget-holders as having a stake in the training process. Therefore, it should come as no surprise to find that these people will all have an interest in the progress of projects.

The project team itself will also need to keep track of how things are progressing. Therefore, we need a system of published target dates and progress tracking. However, once the system is in place, then we need to use it. Periodical progress meetings are a way of identifying problems and seeking solutions.

Computer-based information systems, project management software, company intranets are all available and are probably too sophisticated for the average project. The important point is to keep management informed and one of the biggest complaints from management is that project reports are too detailed. Management is interested in progress so that it can offer support when things go wrong. The most effective support management can offer training design teams is to defer the completion date of a project or to make more resources available. To do that, they need to know what is going on.

Project Control

Following on from the previous topic, as well as monitoring progress, we need a means of control. By control we mean keeping things on track, moving according to plan. This also assumes we have considered contingencies. At an early stage in the project it is worthwhile establishing the risk factors involved. Even in a team of one, a sudden protracted illness could stop things dead in their tracks.

Identifying the risks will probably require a little brainstorming. The obvious project risks are illness in the team, changing jobs and increased tasking. Less obvious are a change of requirement at some stage during the project, unwilling or unhelpful colleagues, withdrawal of funding and support.

Having established a likely list, we can allocate probabilities. Again, this can only be guesswork at this stage. For those risks that we consider to be above a certain threshold, then we need a contingency plan. For example, if a subject matter expert is key to the task, who would be a suitable substitute should something go wrong.

The key control structures are the initial objectives, the schedule or milestones and the budget. These can be reviewed at periodic meetings at which progress is reviewed, risks re-evaluated and resources reassigned as required. As a final warning, in terms of project control, never underestimate the lack of understanding of training as a process among senior managers. Consequently, allow ample time to brief on progress.

Termination

Project termination calls for a clear ending of the activity. We need to establish that the task is now complete. From our earlier discussion of post-course evaluation, we can see that even though the bulk of the course production work has been done and the material handed off to those responsible for course delivery, the need to evaluate will require that project termination be deferred. Furthermore, for Level 3 and 4 evaluation, we said that we needed to allow time for the effects of training to work through into the workplace. As part of our project management we need to make allowances for this deferred termination while, at the same time, keeping control of a process when the bulk of the team have been re-deployed to other tasks.

At some stage, we also need to take stock. In a learning organisation we should be looking at how we can capture expertise so that we can be more effective next time. Therefore, a post-project review should be looking learn from the experience. The management methods should be evaluated and, if necessary, modified. Experience needs to be captured, either on paper or in some digital format. Hard data, such as actual times taken for tasks, resources not used, valuable contacts should be logged. Importantly, the risk management plan can be revisited with a view to refining the list of risks and recalculating the estimated probabilities.

Management of Change

All training design activity will involve a change. In particular, as new technology becomes available, people will want to try things a different way. Not everyone will accept the need. In the pilot training course examined in Chapter 7, one recommendation was that the course be extended by one day. The manager responsible for training in the airline rejected the proposal saying that 'it was good enough for me so why isn't it good enough for today's pilots'. Change management is a complex subject and warrants a book in its own right. However, for now, we will outline some of the key stages in change management which need to be considered as part of the project cycle.

At the project initiation stage, we need to have a clear idea of what we are trying to achieve and, more importantly, who is going to be affected by the changes. As part of our stakeholder analysis, we need to make sure that all the issues have been uncovered and we must make sure that everyone involved has been informed and given the opportunity to make comments.

Change management involves breaking down barriers. Rumours are often a good indication of what misconceptions are already building and what additional communication is needed from you. You should try to identify individuals who are uncomfortable with the planned changes and use their concerns as feedback on the soundness of your design. You should also make sure that management is fully behind the change – and that they make their support known.

As part of your project control mechanism, set targets, hold meetings, visit interested parties on their territory, keep track of progress and make sure that you have facts and figures to hand, if necessary, to support your actions. Never underestimate the need to communicate with absolutely everyone affected.

As far as possible, changes should be field-tested, preferably by groups who are supportive of the change. We were once involved in trying out a new approach to pilot training in a flight simulator. Two crews were given the task of testing the revised training method, one of whom was against the project. Needless to say, the crew who did not support the proposal managed to undermine the trial such that no useful data was gained. As part of the small-scale trials, try to anticipate which parts of the organisation will be affected and actively seek information about unexpected effects. Be prepared to change plans and always explain why such changes are taking place.

At the end of it all, gather information relevant to the change management process in the same way that you attempted to record lessons learnt from the main training design activity. It is true to say that any training activity is also a change agent in that trainees return to the workplace with altered views of their jobs. This, in turn, needs management.

Conclusion

In this chapter we have offered a brief overview of the stages of project management with some examples of how it can all go wrong. There are many similarities between some of the stages and activity normally associated with training design. However, our motive for including a separate chapter on the subject is simply that, all too often, training design projects fail for lack of effective management. We have already seen that some aspects of the design cycle, such as testing and evaluation, need to be considered at the outset and not as afterthoughts. For the same reason, effective management structures need to be in place from the start of the project. On the other hand, we have worked on projects where the management got in the way of course production. Team members spent all of their time drafting design specifications and updating flowcharts. We need to strike a sensible balance but, at the same time, we cannot afford to forget that we are still supposed to be in control of the process.

Case Study

You have been assigned to a team responsible for introducing a cabin simulator for use in safety training. A suitable area within your building has been identified for locating the device, which has already been ordered and will be delivered in 6 months time. It is anticipated that joint classes of flight deck and cabin crew will be trained in a group size of 14. The course, which is currently being conducted by an outside company, will last two days. No other training is being conducted in the building at present although if this first project is a success, then additional courses, such as customer care and basic induction training for cabin crew, may be brought in-house. Your task is to identify the key project management aspects we will need to consider to make sure that the exercise is completed with the minimum of fuss.

9 Quality Management

Introduction

In this chapter we will consider the application of the principles of Total Quality Management (TQM) to training design and training management. The concept of TQM has been influential in manufacturing for many years and has started to take hold in the service industries. Training clearly falls into this latter category. Through our examination of TQM we will be developing a model of a training system which is constantly renewing itself; perpetual training design. We will look at the how existing models of TQM apply to training and see what systems we need to have in place in order to achieve our goal. We will also look at some of the weaknesses in the TQM model that we need to be aware of. In so doing, we will be revisiting some of the themes of earlier chapters. Finally, we will look at some initiatives in performance monitoring that may be of use to us in the future. At the end of this chapter you will have:

- Considered the application of a TQM model to the training process.
- Identified some structures needed to support the TQM principles.
- Considered some weaknesses in the TQM model.
- Evaluated recent training initiatives in terms of TQM.

The TQM Model

Total Quality Management has its roots in engineering and statistical process control. It sees activity in terms of a production process. Thus, there are inputs that are processed to achieve a planned output. The output needs to be measured against a standard, or product specification, and action must be taken if output fails to match the standard. The process will have been designed to convert inputs to output in the most economical fashion. Feedback loops control activity at every stage. A typical TQM model is illustrated in Figure 9.1.

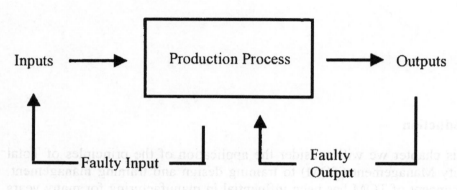

Figure 9.1
A Simplified Total Quality System

In a training system the inputs, in the sense of raw materials, are clearly the students requiring training. But inputs can also be the teaching activities which aid the student's progress, the instructors, the type of aircraft or simulator used to accomplish a goal and so on. In fact, any resource that is used in the training process can be considered as an input and, therefore, be required to meet a quality standard.

The output from our system could be the trained person or it could be the training scheme as a whole. For example, quite often it is the 'course' that is offered to prospective customers and therefore is an entity in its own right.

The 'production process' element of the model is, principally, the scheme of instruction. However, we can also include the training environment, supporting activities, pre-and post-course activities as part of the production process if they form an integral part of the whole training activity. For example, many airlines run senior management courses or CRM courses at out-of-town hotels in order to isolate delegates from the demands of their daily jobs and, in some cases, to contribute to a feeling of 'worth'. Therefore, even beach resort hotels can be considered a part of our production process!

So, although the exact comparison between manufacturing and the training of airline personnel is difficult, it is not impossible. The TQM model does allow us to identify areas of interest that will have an impact on the efficiency of the training process. We will examine the critical aspects of the model in more detail later. First, we need to examine what we mean by 'Quality'.

Some Fundamental Principles of Quality

However you choose to define 'quality', there are some guiding principles that underpin the concept. First, like safety, quality must start at the top of the organisation and cascade down. When asked to devise a quality system for a pilot training school we pointed out that periodic review meetings would be needed and it would be advisable if the Chief Executive chaired these. On being told that the Chief Executive was too busy to get involved, we realised that quality was not something the organisation aspired to. In addition to starting at the top, quality is something that all levels of management need to buy in to.

The second guiding principle is that quality revolves around customer relationships. We are all either the customer of someone else or have customers of our own. We should expect the best service as customers and offer the best service to our customers. In any organisation there are internal customers, i.e. other departments we support, as well as external customers, i.e. those who pay our wages. Quality is all about optimising these customer transactions so that everyone goes away satisfied. For example, consider the case of a pilot ground training school that was having difficulties getting students through their final licensing exams. At the time, the aviation industry was in recession and few airlines were recruiting. As a result, the prestigious airline cadet schemes, on which the school had based its reputation, were few and far between. The bulk of their students were the self-improvers, those dedicated individuals who work and save in order to climb up the pilot training ladder a rung at a time. Because of the focus on airlines, the school offered a poor service to the self-improvers and, consequently, did considerable damage to the school's reputation. Here, the customer was the self-improver attending a short course of two or four weeks rather than the 16 month full course. Because of the lack of concern for the customer, and the failure to spot that, given the prevailing economic situation, this was the company's main revenue stream at this stage, the quality of the output, the course offered for sale, suffered.

Finally, quality is about tracking inputs and outputs. If we fail to understand either, our system will fail. We have talked elsewhere about input and output standards within training systems. The important point is that, as part of TQM, we are required to actively measure both and use the data for management purposes; in effect, we take control of the process.

Stakeholders in Quality

In Chapter 2 we introduced the concept of stakeholders, all of whom had an interest in training. In the same way, developing quality systems requires us to recognise the needs of stakeholders. If we stay with the training of commercial pilots for a moment, we have already seen that the TQM model incorporates the idea of internal and external customers. The external customer in this case could be the self-sponsored student (the self-improver mentioned earlier) or a contracting air carrier. The internal customers are the different parts of the training organisation that participate in the process. Thus, ground instruction often takes place away from the flight instruction. Single-engine and multi-engine training can be given by different groups of instructors. Students may still be attending ground instruction whilst progressing through the flying phase. Progress in flying may be held up by the need to re-sit a part of ground school. The important point is that an internal marketplace does exist in which students are passed from one production unit to another.

We listed some of the key stakeholders in training in Chapter 2. It is unfortunate that stakeholders do not necessarily share a *common* interest. For example, self-sponsored students want to feel that their investment has been worthwhile, that they have a high probability of achieving their licences at the first attempt and that they have a good chance of employment at the end of the course. They, possibly, want to have a good time in the process. An air carrier operating a sponsorship scheme wants to ensure that those starting training are still there at the end of the course and that they can be productively employed with the minimum of additional training. On the other hand, shareholders in the training company want the maximum return on their investment. Given that ground instruction typically takes most time whilst contributing little in the way of revenue, shareholders will want to see minimal time in the classroom and maximum utilisation of the aircraft. They, too, will want to see students pass ground exams first time but are probably less concerned about the need for additional flight instruction to pass a test. Regulators are concerned that trained pilots meet the standard required to maintain the safe operation of the national aviation system. Through their flight crew licensing office, the Regulators have devised a ground curriculum that is described, in many cases, in terms of numbers of classroom hours. But, we have already seen that ground instruction does not make money and so schools find ways to shave off hours whilst still making sure that students stand a chance of passing the written test. In short, the players in the game have different, and sometimes incompatible, goals.

Quality, like training, is a delicate balancing act that involves meeting the, often conflicting, needs of a range of interested parties. As training designers, we are mainly concerned with making sure that the production process is as effective as possible. In the same way that we considered evaluation as something to be considered at the design stage, as we put our training system together we can start to think about the structural elements we would need in place if quality management was to be applied once the course goes live.

In addition to giving advanced warning of what to consider when designing training, stakeholder analysis can shed light on why well-designed systems can still fail or, perhaps perversely, why it is almost impossible to design a Total Quality system. After all, quality is largely in the eye of the beholder.

We have said that a TQM model of training identifies inputs, processes and outputs. It can identify choke points and limiting factors in the process. But here we need to sound a note of caution. A TQM system it is blind to motive. It can tell you what is going on but not necessarily why. On that note, we will now examine a TQM system in more detail. In addition to the elements already discussed, the components that typically form part of a TQM system are listed in Table 9.1.

What do we Already Have?

Although set in different jargon, we can see already that many of the components of a quality system exist within our systematic approach to training design. For a start, we spent most of the earlier part of this book looking at the need for, and ways of defining, the Product Specification or Course Syllabus.

Most training systems have a way of controlling and tracking inputs. For external applicants we have selection systems and for internal applicants the choice is usually made for us; we are running training because the people need to be trained. However, although it is beyond the scope of this book, methods of selection can play an important part in the success of our training system. The US Armed Forces did a trial looking at operators of Patriot guided missiles. They applied a range of selection tests to the applicants and then, on the basis of the results, divided them into groups. They then trained the groups to operate the missile system. They found that the trainees who had been in the top group, on the basis of the selection tests, achieved the proficiency standard with one less live-missile firing than those in the next group down. Given that each missile cost something in the region of US$ 1 million, we can see that control of inputs – the trainees – can have a significant effect. Our experience of airline recruitment is that it can often be haphazard and is usually undertaken by enthusiastic but unskilled staff. The relationship between selection methods and success in training is a question rarely asked.

Table 9.1
Components of a Quality System

- A statement of top management's commitment to quality.
- A defined product standard.
- A system to control and track inputs.
- A system to detect, control and re-work non-conformities.
- A system to prevent further non-conformities.
- A method of verification
- Independent audits
- Methods of process control, including course administration.
- Post-production functions, such as after-sales follow-up of output.
- A system of documentation control

The method by which we take the raw materials and turn them into finished products is the primary focus of TQM. A flawed product specification, or syllabus, results in us doing the wrong job but a flawed production process, or curriculum, means that we are doing the job wrong.

If we start by considering production as a standardised activity, and that detecting deviations from the standard is the essence of TQM, then we can immediately see differences between manufacturing and training. Consider basic pilot training. The relationship between instructor and student pilot tends to be a highly personal one-to-one relationship with no two students receiving the exact same training under the same conditions of weather, traffic density and so on. Flying training is, possibly, more like small batch production where the process is altered from one run to the next to meet the individual requirements of the particular customer.

However, we need to be wary of trying to draw parallels between, in this case, pilot training and engineering processes. At this point we need to recall the earlier discussion of organisational factors and their influence on training systems. National approaches to pilot training lie along a continuum that ranges from vocational, at one extreme, to educational at the other. Thus, the UK model is generally divorced from the national educational system and concentrates on meeting the needs of ICAO Annex 1 whilst getting pilots productive as fast as possible. At the other extreme, Russia, and university degree-based courses in the States and elsewhere, contain large blocks of instruction which meet curriculum goals decreed by non-aviation bodies. The first two years of the Russian system was designed to produce a trained engineer. Achieving 'engineer' status guaranteed a set minimum wage and offered a fall-back if a student failed flying training. In the US system, university accreditation bodies require minimum standards of academic input in order to maintain standards. Variations on these themes can be identified. In effect, the best we can hope for is a standardised process mediated by stakeholder demands.

At the macro level, then, the design of the production process reflects a host of influences which the application of the TQM model difficult. However, at the micro level we also find anomalies. Some things are taught in groundschool because they are needed for the exam, not because they are needed for subsequent airborne exercises. Regulators tend to add items to the syllabus at a faster rate than they remove them. In individual subjects the treatment can differ between the abstract, theoretical ground training approach and that of the more pragmatic, operationally-oriented flight instructors. Thus, in its strictest sense, TQM would fail when applied to training. However, provided we accommodate these factors, there is still much to be gained by working within the discipline of the quality concept.

We now move on to another fundamental principle of quality management; measurement of the process. Of course, measurement was covered in Chapter 5 when we looked at methods of testing but, at this stage, we just want to offer a few observations on testing in the light of

TQM. It was said earlier that one of the roots of TQM was statistical process control. This requires the process to be monitored in a quantitative way such that deviations from the plan can be identified and rectified. Implicit in all of this is that the process is capable of measurement in some meaningful way.

If we continue with basic pilot training as an example, the main measures used in pilot training are hours flown, accuracy of manoeuvre and achievement in ground examinations. Some of these may be seen as continuous variables. That is, we are sampling activity and the variable gives us an indication of changes in performance. So, the variation about the required altitude of a student trying to maintain level flight can be considered a continuous variable. Values will change in relation to changes in performance and, over time, students' scores will show an overall move in some required direction. Others measures are discrete variables in that they represent a snapshot of performance in time. Thus, exam scores are discrete variables in that give us a value for a performance on a particular day.

We make a number of assumptions about ability on the basis of these measures without necessarily questioning the true value of the measures. For example, we said that hours flown is an indicator of performance; the more hours you have the greater your level of experience is deemed to be. However, there has long been a great dissatisfaction with total hours as an indicator of ability. The link between hours accumulation and proficiency makes intuitive sense but is difficult to establish. For example, we once examined the records of UK Royal Air Force C-130 Hercules First Officers. After six months on their first squadron, First Officers were expected to achieve the next-higher proficiency grading which would increase their flexibility of employment. The numbers of First Officers failing to upgrade at their first attempt was seen to indicate a decline in the input standard to the training system. However, an examination of log-books revealed a different picture. Squadrons were required to provide complete crews for recurrent training. Because of scheduling constraints, a First Officer could be put up for re-categorisation after between four and seven months on the squadron. It was found that First Officers who had achieved a certain threshold value of hours on the squadron, or who had served at least six months, rarely failed to upgrade. Those who did not match one or other of the criteria were likely to fail. So, it was not a decline in quality but rather poor management that was causing the problem. Furthermore, hours on type was, in part, a predictor of success, but it was still not possible to say why gaining a certain number of hours would almost guarantee a pass.

Whenever we set out to measure something, we need to choose indicators which are valid and reliable, in the same way that we said that testing needed to be valid and reliable. Lets look at validity in the context of quality management. In the C-130 example, it was assumed that the decline in the percentage of pilots upgrading was an indicator of the standard of First Officer. Discussions had already begun with basic training organisations about the product of their system and yet the indicator was not valid. In fact, it was better indicator of First Officer career management than it was of basic ability. To take another example, pilots are required to pass a series of ground examinations before they can be awarded a license. In an examination of UK CAA exams, it was found that 80% of questions required simple recall of facts. The point here is what do we believe success in the exam is telling us about the standard of pilot entering the industry? Performance in the real world requires pilots to apply knowledge to solve problems. We saw in Chapter 4 that it is easier to design questions which require recall from memory than it is to design items which test the application of knowledge. However, in this context, is an exam that is predominantly a test of memory a valid indicator of possible future performance?

If we now move on to reliability, we have seen that this concept deals with the extent to which we can have faith in the results of our measure. Thus, if I get 90% in the Air Law exam, does that mean that I know more than about Air Law than someone who gets 60%? What happens if we both resit the exam in a weeks time and I only get 75% whilst they get 80%? In TQM terms, measurement must be directly linked to the production process such that activity can be monitored and modified if it fails to contribute to the desired outcomes. Ground examination results are probably hopeless as performance indicators, both in terms of the true ability of the students and the effectiveness of the learning system. To reinforce this view we cite the example of a European carrier's experience of using CBT as part of its ab-initio programme. An examination of the computerised student records showed that some students not only failed to complete elements of the course but they also managed to get less than 50% on the computer-administered tests. Yet those same students passed their Aviation Authority ground exams with almost perfect scores.

This discussion of performance measurement reiterates the doubts raised about the traditional measures of student achievement in the earlier discussion of testing. However, the fact remains that measurement is crucial to training. This is increasingly the case in aviation as the concept of 'at-risk' pilots grows and accidents are increasingly traced back to poor performance in training or in earlier employment.

What is Missing?

Having looked at which of the key components of a quality system already exist, we will now consider what we need to add in order to make the system work. Probably the first observation to be made, and which can be leveled at training as a whole, is that most systems lack the rigour necessary to measure up to the discipline of TQM. Much training is done on a minimal budget by staff who have received little or no training in instructional techniques. The 747 taking off from Los Angeles, cited in the Introduction, was the victim of a training system which lacked any concept of quality management. After a Beech 1900 crashed into the sea of Rhode Island in 1991, the US National Transportation Safety Board found that the airline's Director of Training was not qualified on the aircraft type. Furthermore, no systems had been established to monitor or standardise training and that no one seemed to be in charge of monitoring training on the Beech.[26] On a more mundane level, we were told of a CRM course in the UK that used classroom hand-outs collected during some other training provider's course and which retained all the identification marks, logos, titles, etc of the originator. Without levels of discipline previously absent from a significant proportion of the training industry, any investment in quality would be wasted.

The next key element of the system that needs to be enhanced – or even developed – is the process by which non-conformities are dealt with. Of course, non-conformities are principally those students who fail the course. In many cases, repeating elements of the course or re-sitting exams are the only strategies adopted. Unfortunately, if the student failed to learn effectively on the first pass, simple repetition is unlikely to solve the problem. At this stage we need to return to the analysis of learning needs. Having based our course design around some estimation of the most probable solution in terms of student learning needs, our remediation systems need to reflect the less-common alternative learning strategies demonstrated by course members.

Non-conformities also include any structural elements which are under-performing, such as the course itself. Therefore, evaluation starts to feature strongly in our approach to training design.

A fully-functioning TQM system also requires some degree of independent auditing. We will look at a form of surrogate auditing later in this chapter. However, many airlines have developed auditing departments and, in JAA States, the Quality plans that form a part of the company operation require auditing mechanisms. Clearly, this is in addition to the course evaluation procedures discussed in Chapter 6.

Finally, and continuing the theme of evaluation and auditing, Quality requires closer relationships to be developed with customers such that a process of continuous improvement can be adopted. In many cases, airline trainers are sufficiently close to the operational work-face to allow them to keep track of developments in the real world. However, it may require a system of line visits, or rostering trainers to fly occasionally, to ensure effective communications between the training supplier and the operational customer. This is a particular problem where training is offered by a third party.

Some Problems with Quality

Having an over-arching quality system will undoubtedly bring benefits to an organisation. From the discussion so far, a quality structure can be installed by strengthening the activities which form part of the training design cycle and making certain other additions, which do not need to be extensive. However, there is a cost to be paid.

The direct cost of quality can be measured in terms of the additional bureaucracy and maintenance required in order to ensure that all training resources are tracked and maintained to current standards. Some organisations are requiring accreditation of third party providers by Quality Standards organisations, which costs money. Independent auditing requires additional staff or sub-contractors. Even the use of internal staff who are working to a quality checklist still represent effort diverted from their primary tasks.

Finally, the selection of realistic performance indicators can often be more difficult than first imagined. Just as in the case of testing students, what can easily be measured may not necessarily reflect any significant aspect of training system performance. This is especially the case if the training organisation has no control over the variable being measure. For example, one school we looked at used the ratio of students trained to the total students planned for training at the start of the financial year as one of the performance indicators. However, as the school was not responsible for recruiting students in the first place, they had no control over who turned up for training. The statistic generated reflected the efficiency of the recruiting system but said nothing about the quality of training offered. The selection of performance indicators which allow some degree of effective management of the training process requires an investment of time and effort and also, probably, the application of information technology, which costs money.

Benchmarking

In the previous section we talked about the need for independent auditing as part of a TQM system. In this section we want to look at a method of auditing by default: Benchmarking. The concept of benchmarking is quite simple. We set up a standard and compare our performance against the standard. In this case, the standard is the performance of other organisations either in the same line of business as ourselves or who use processes that are analogous to ours and from which we think we can learn some lessons. Of course, we run the risk of sharing valuable information with our competitors when benchmarking with organisations doing the same job as ourselves, but this is simply needs managing and need not be a bar to effective benchmarking. As a process, benchmarking has been described as 'Industrial Tourism'. In order to take full advantage of the activity, we need to plan.

Step One in a benchmarking project is to identify the component of the system to be checked. It is probably unrealistic to make comparisons between complete training systems. Therefore we need a focus. It could be an investigation of approaches to security training or the selection of new entrant cabin crew.

Step Two requires us to find an organisation that does the same job as us or, as we said earlier, an organisation that does something that is analogous. For example, many airlines recruit large numbers of short-term cabin crew for the summer peak season. We may want to look at an organisation which has a similar demand for large numbers of short term employees on a seasonal basis, such as tour guides, periodic sporting events like the World Cup Football Competition, and so on. As part of our planning we need to brainstorm possible benchmarks.

Having narrowed our list down to a reasonable number of probable sources of information, Step Three involves negotiating access. This will no doubt involve return visit, but the whole point of the exercise is to share information and learn from the experience. If we have any worries about letting others see how we do things, then we are probably in such a bad state that benchmarking is the least of our worries!

Having arranged our visit, in Step Four we need to structure the collection of data. We have already identified the area under investigation, so we need a few well-planned questions to get at the information that will be of benefit to us. We need to look closely at the procedures being used and see how ours can be improved. As part of the data gathering we need some performance indicators in order to permit sensible objective comparisons between the organisations being examined. In a training system we can look at:

> Measures of Input – e.g. Staff ratios, Per capita expenditure, Overheads.
> Measures of Output - e.g. Time to proficiency, Cost per trainee, Customer satisfaction

Finally in Step Five, on our return, we do something! We need a commitment to implement any changes that might be suggested as a result of the exercise. And here we need to reflect on the problems of change management discussed in Chapter 8.

Of course, benchmarking is not an independent audit in the sense of the TQM model but it does allow some examination of how we are doing in comparison with others. In the next section we want to look at how, using technology, we can start to build training systems which are self-auditing and continually updating.

Perpetual Training Design

Advances in technology now allow us to gather real-time data about some aspects of crew performance. Linked to cheap yet powerful computers, we can start to consider training systems that continually renew themselves. We want to build up a model of such a system as it might apply to airline pilots. However, the principles could be adapted to any occupational group. At the core of the system is an accurate task description and a data collection method.

For many years aircraft have been able to store data about particular performance parameters. This is the information extracted from the 'Black Box' as part of an accident investigation. As aircraft have become more complex and more 'electronic', ever-increasing quantities of data have been generated and stored. A modern Airbus aircraft has over 1000 bits of information being recorded about how different aircraft systems are performing and how the aircraft itself is progressing on its journey. For 30 years many of the larger airlines have been using this data for performance monitoring. Initially, the information was used to check on how systems such as engines were functioning and to try to anticipate any problems that might be predicted by changes in parameter values. More recently, pilot operation of the aircraft has become a focus of attention. So, if a pilot completes a task that, in some way, contravenes a limitation, the system can detect the event and the pilot can be counseled accordingly. Other operational problems, such as fuel consumption and brake wear can also be monitored and procedures changed to allow the airline to save money.

Whereas, in the past, data monitoring takes place after a significant event, such as a heavy landing or an accident, or as a result of some maintenance management scheme, our perpetual training design model requires aircraft data to be routinely downloaded and examined. In the USA this policy is being introduced under the auspices of the Flight Operations Quality Assurance (FOQA) programme. In the UK we use the acronym OFDM but it means the same thing. By using automated search procedures and batch processing by computers we can rapidly analyse the masses of data generated each day by the average airline. But what we do now is not to look for exceedences or faults but to monitor crew performance against some ideal standard.

The standard is derived from an accurate task analysis. Again, some flying organisations, typically the military, have defined the flying task as part of the basic training programmes for many years. It allows a degree of standardisation of the training and lets student pilots monitor their own progress. In the US, the Advanced Qualification Programme (AQP), set up in the early 1990s, provided for airlines to develop flexible training programmes based on an anlalysis of their line flying task. With the task analysis, we can now start to look for reliable performance indicators which allow us to discriminate between effective and less effective pilots. This is not as easy as it may sound. After all, piloting an aircraft is a collaborative task undertaken in a range of conditions.

Although the task analysis and the data gathering are the heart of the system, they would not achieve our goal of a self-regulating system. For a start, the stored data only measures a sub-set of the total skill set of the pilots. Only those skills that require the pilot to interface with the machine can be measured. However, we have other sources of data. Pilots make use of simulators for training and the output from these is also digital. We can now start to compare performance in the simulator with that in the aircraft to see if there are any anomalies. Pilots also have line checks during which their performance is observed by a skilled assessor. By collecting this observational data in a suitable format, say by making entries into a spreadsheet on a hand-held computer, we can start to develop a more comprehensive picture of pilot performance. By storing data for all pilots in a searchable format we can start to see any gaps in real-world performance which may indicate a problem with our training.

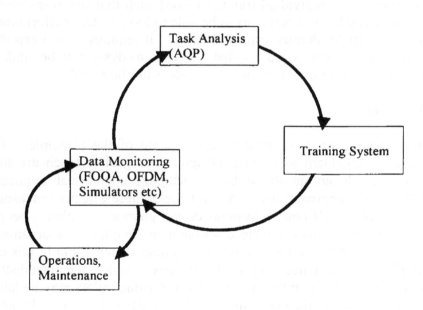

Figure 9.2
Self-regulating Training

But it doesn't end there. The task analysis represents a snapshot in time of how that airline operated. Because it is an artifact of the analysis process, it is subject to all of the weaknesses described in Chapter 4. So we need a way of refining the analysis. In part, the data from the aircraft, analysed in terms of fleet performance, will give us some indication of a problem, as discussed above. However, pilots submit a variety of reports relating to daily problems: voyage reports, safety reports, anonymous confidential reports, mandatory occurrence reports and so on. These are all sources of data which, when aggregated, could indicate training problems. We are now moving into the realms of what is fashionably called knowledge engineering. Most airlines still see much of this type of information as something to be hidden, to be ignored or something for which a damage limitation response needs to be generated. In fact, when handled correctly, it is a rich source of information for training designers.

So, our self-perpetuating training system needs to be based on a task analysis, a means of data gathering and, finally, a means of data integration. It requires as many samples of behaviour as possible to be incorporated into individual training record such that our measurement is valid and reliable. It requires us to be able to look at fleet performance as a whole in order to identify gaps and trends. It requires us to keep track of changes in the real world so that the task analysis can be updated as required. We have tried to reflect these ideas in Figure 9.2.

Conclusion

In this chapter we have tried to apply some of the principles of Total Quality Management to training design. Our goal is to ensure that our training is relevant in the light of changing operational requirements. Implicit in the approach was a desire to seek more effective measurement of performance. Of course, systems do not guarantee quality – people do. We also need to touch on the discussion in Chapter 2 of organisational factors. One of the barriers to the ideal system sketched out in this chapter is employee acceptance. Pilot unions have resisted the introduction of routine data analysis on the grounds that individual pilots may be laid open to legal action should data fall into the hands of lawyers. In addition, management might use data as a way of taking punitive action against individuals. These are all valid stakeholder views that need to be addressed. For now, we are simply outlining possibilities.

Just as the previous chapter highlighted similarities between project management and training design activities, so this chapter has seen commonality between training design and quality management. This should come as no surprise to us because, at the end of the day, training design is all about the application of management principles to the process of identifying training requirements and developing training solutions. We come down to the same set of questions:

- 'Are we doing the job correctly?'
- 'Do we continue to do the job correctly?'
- 'Have we done the job more correctly than necessary?'
- 'Could we do the job more consistently and on target?'

We hope that this book has shed some light on the problems involved in seeking answers to those questions.

Case Study

You have been called to your line manager's office. She hands you this checklist of items she has taken from an article on TQM. She wants you to go back to your office and look at how many of these things are already in existence, in one form or another, in your training organisation. She also wants you to consider which items could be introduced at little or no cost. Finally, list those items that would require considerable effort. With that done, she wants you to identify the key items on the list, by which she means those without which a quality system would most likely fail. You amble back to your office and start work. Whilst you tackle the task, bear in mind either your current job or the last training organisation you worked in. You may want to use the list to compare organisations, or training sections in your own organisation, and to see if it is possible to identify qualitative differences on the basis of the list:

Management responsibility
Quality systems principles
Internal audit
Contracts
Design control
Input control
Process control (includes administration)
Production
Materials and traceability
Inspection and testing

Control of inspection and testing equipment
Non-conformity
Post-production functions
After-sales (follow-up of output)
Documentation control.

Notes

1. Bureau of Air Safety Investigation Report No 9403038 dated September 1996. "Boeing 747-312 VH-INH, Sydney (Kingsford-Smith) Airport, New South Wales, 19 October 1994".
2. Incident reported in the NASA-run Anonymous Safety Reporting System.
3. *Air Safety Week* Vol. 13, No. 15 dated 12 April 1999.
4. *Air Transport World*, October 1995 pp. 51-57.
5. Kolb, D.A. (1984) *Experiential Learning*, Prentice-Hall, Englewood Cliffs, N.Y.
6. Honey, P and Mumford, A. (1992) *The Manual of Learning Styles*. Contact Dr Peter Honey, Ardingly House, 10 Linden Avenue, Maidenhead, Berkshire, SL6 6HB, UK.
7. Telfer, R.A. and Moore P.J. (1995) 'Learning, Instruction and Organisation in Aviation', Paper presented at the Eighth International Symposium on Aviation Psychology, Columbus, Ohio.
8. Henley, I. (1995) 'The Quality of Flight Instructor Training in Australia and Canada', *Aviation Psychology: Training and Selection*, Proceedings of the 21st Conference of the European Association for AviationPsychology. Eds, Johnston, N., Fuller, R. and McDonald, N.
9. Hunt, L (1995) 'Effective Learning Strategies of ab initio Pilots', *Aviation Psychology: Training and Selection*, Proceedings of the 21st Conference of the European Association for Aviation Psychology. Eds, Johnston, N., Fuller, R. and McDonald, N.
10. Bloom, B.S., Engelhart, M.D., Furst, E.J., Hill, W.H., & Krathwohl, D.R. (1956) *Taxonomy of Educational Objectives, Handbook I, Cognitive Domain*, New York. Longmans, Green.

11. Norman, D.A. (1988) *The Psychology of Everyday Things*, New York, Doubleday.
12. Hackman, R. and Oldman, G. (1975) 'Development of the Job Diagnostic Survey', *Journal of Applied Psychology* 60 (1) pp 159-170.
13. For advice on a range of task analysis methods see Kirwan, B. and Ainsworth, L.K., (1992) *A Guide to Task Analysis*, London: Taylor and Francis.
14. Reason, J. (1990) *Human Error*, Cambridge University Press.
15. FAA Advisory Circular AC-120-54A dated December 1998. 'Advanced Qualification Program'.
16. National Vocational Qualifications. Piloting Transport Aircraft, Aviation Training Association, High Wycombe, UK.
17. Feggetter A.J.W., McIntyre H.M. and Mortenson, L. (1987) 'A Task Analysis of Engine Off Landings', Army Personnel Research Establishment Report No 87R011, Farnborough.
18. Stevens, A., Collins, A. and Goldin, S.E., 'Misconceptions in Students' Understanding'.
19. de Corte, E. (1990) 'Learning with New Information Technologies in Schools: Perspectives from the Psychology of Learning and Instruction', *Journal of Computer Assisted Learning*, (6) pp. 69-87.
20. Gopher, D, Weil, M and Bareket, T. (1994) 'Transfer of Skill from a Computer Game Trainer to Flight', *Human Factors*, 36(3) pp. 387-405.
21. Horgan, J. (1992) *Common Principles for the Assessment of the Cognitive Results of Training*, Commission of the European Communities.
22. 'School Testing – Private, keep out', *The Economist*, 18 June 1994.
23. Kedem, A. and Smilansky, J. (1985) 'Design and Production of Training Packages to Accompany the Development of Fighter Aircraft', Paper presented to the Third Annual US Interservice-Industry Training Equipment Conference, Florida.
24. Wrightman, D. (1983) *Part-task Training Strategies in Simulated Carrier Landing Final Approach Training*. Naval Training Eqipment Centre Technical Report: NAVTRAEQUIPCEN IH-347.
25. Kirkpatrick, D.L. (1959) 'Techniques for evaluating training programmes', *Journal of the American Society of Training Directors*, 13 pp. 3-9, 21-26; 14 pp. 13-18, 28-32.

26. NTSB Accident Report NYC-92-FA-053. *Beech 1900C, N811BE crashed near Block Island, Rhode Island on 28 December 1991.*

Index

For Product Safety Concerns and Information please contact our
EU representative GPSR@taylorandfrancis.com Taylor & Francis
Verlag GmbH, Kaufingerstraße 24, 80331 München, Germany